CLIMATE
CONUNDRUM

THE **AGENDAS** AND **FORCES** AT PLAY

CLIMATE
CONUNDRUM

THE **AGENDAS** AND **FORCES** AT PLAY

BALVINDER RUBY

Climate Conundrum - The Agendas and Forces at Play
Balvinder Ruby

Published by White Falcon Publishing
Chandigarh, India

The contents of this book have been certified and timestamped
on the Gnosis blockchain as a permanent proof of existence.
Scan the QR code or visit the URL given on the back cover
to verify the blockchain certification for this book.

ISBN - 978-1-63640-976-4

Dedication

This book is dedicated to my beloved father, Lachhman Singh Dhaliwal, and mother, Kirpal Kaur Tiwana. Your unwavering love, support, and encouragement have been the guiding force in my life.

Your selflessness and sacrifice have made it possible for me to pursue my dreams and passions, and I am forever grateful for the values and principles you have instilled in me.

You have shown me the importance of hard work, determination, and compassion, and I hope that this book serves as a testament to the values that you have taught me.

Thank you for always believing in me and being my pillar of strength. This book is a tribute to your love and sacrifice, and I dedicate it to you with all my heart.

Contents

List of Illustrations

Acknowledgements

I would like to express my deepest gratitude to all those who have supported me throughout the writing of this book; without their encouragement, feedback, and assistance, this work would not have been possible.

First and foremost, I would like to thank my family for their unwavering love and support, for always believing in me, and for giving me a free hand to pursue my pursuits. Their encouragement has been instrumental in helping me pursue my passion for investigating, researching, and commenting on important issues having a bearing on society.

I would also like to express my appreciation to my friends and colleagues who have provided me with valuable feedback and insights on this work. Their input has been invaluable in shaping my ideas and improving the quality of my writing.

I am also grateful to numerous scientists, researchers, and experts who have devoted their lives to studying the subject of climate change and its impacts; their tireless work has provided the foundation for this book and helped raise awareness about this critical issue.

I would like to express my heartfelt appreciation to Dr. Jiwan Dhillon, PhD, former Director of Estimates Coordination at the Defence Materiel Organisation, Government of Australia, for his invaluable contribution in critically evaluating the manuscript. His keen eye and expertise in identifying structural and conceptual mistakes and inconsistencies, as well as any glaring gaps in the reasoning and presentation of the subject matter, have significantly enhanced the overall quality of the work. I am sincerely grateful for his insightful feedback, which has played a crucial role in refining and strengthening the manuscript. His dedication and meticulous attention to detail have been instrumental in shaping the final outcome, and I am truly grateful for his valuable assistance throughout the process.

Finally, I thank my publisher and editor for their guidance, support, and patience throughout the publishing process. Their expertise and dedication have helped transform my vision into a reality.

I hope this book will help raise awareness, answer some of the questions, and enable people to ask more pertinent questions and understand the issue of climate change.

Balvinder Ruby
Sydney, June 2023

Preface

The world is currently seized by the issue of climate change. As an earth scientist and researcher, I am keenly aware of the scientific consensus that attributes climate warming to human activities, particularly the combustion of fossil fuels. This consensus, heavily relied upon by the Intergovernmental Panel on Climate Change (IPCC), forms a crucial foundation in our understanding of the causes and potential consequences of climate change. However, it is important to recognize that this consensus, like any other, is not static and is subject to revision as new evidence emerges and scientific inquiry progresses.

In my book, "Climate Conundrum, The Agendas and Forces at Play," I embark on a nuanced exploration of the politics and science underlying climate change. Leveraging my background as an earth scientist and researcher, I delve into the challenges faced in reaching a consensus on the causes of climate change. I also examine the impact of policies such as carbon pricing, green deals, and green finance, and shed light on the motivations driving the actions of corporations and governments.

Through this book, I aim to provide valuable insights into the intricate issue of climate change. I emphasize the significance of open debates, constructive criticism, and thorough scientific research. It is crucial not to employ the consensus model as a means to stifle legitimate scientific inquiry; rather, we should foster an environment of open dialogue and constructive criticism.

I believe this book is essential reading for anyone seeking a deeper understanding of the complexities surrounding climate change, as well as the politics and policies intertwined with it. My unique perspective as an earth scientist and researcher contributes a valuable voice to the ongoing conversation about climate change. It underscores the urgent need for decisive action to address this critical issue, and highlights the importance of informed discussions and evidence-based solutions.

CHAPTER 1

The Hype

Without a doubt, the weather is a common choice as a conversation starter between strangers. The climate is the result of long-term weather trends in a certain area. It would appear that climate change is a topic that can interest everybody, and sensationalising it and the environment by creating a sense of uncertainty and anxiety is a surefire way to keep people interested. Regardless of their knowledge of science in general or climate science in particular, almost everyone seems to be hyped about the subject and is eager to join the movement as an activist or representative of the climate brigade.

On scanning the literature on climate change, it is found that there is a lot of noise without substance. The name "climate scientists" is often mentioned, but the scientists are nowhere to be seen in the debate. Novices, activists, and politicians have hijacked the climate debate, and the scientists themselves are few and far between. Most of the debaters presume themselves to be specialists on the subject and start the chat with the assumption that the climate is changing for the worse and that it is because of human activity and carbon dioxide.

These opinions appear to be formed based on the public perception and narrative shaped by the media and politicians. They believe that the issue of climate science is settled and that humans play a role in causing the Earth to warm. If you ask them how they arrived at this conclusion, they would have no clue because this is the only perception making the rounds, which is a media creation. This question itself is the subject of research on how this perception was generated.

> *"Whether dangerous human-caused climate change is a fact or a fabrication depends on who you choose to believe." Many of us line up somewhere between probable and possible on this spectrum.* John Roscam, Australian Financial Review, 2006

Most of the literature about climate change revolves around who believes in it and who does not, and what percentage of the public or the research papers agree with it, not about the actual phenomenon of climate change itself. For example, a policy document, "The Politics of Climate," by Cary Funk of the Pew Research Centre, focuses on and discusses the views of political outfits on climate change rather than the issue of climate change. Most of the literature is more about the politics of climate change than the science behind it. The issue of climate change has become a means to peddle a hidden agenda to create and encash fear.

It is not difficult to imagine who is behind this extreme hype and hysteria, who is in charge of crafting and establishing a narrative so passionately based on post-truth, and who has a grudge to stifle. The people, whether scientists or social scientists, who stand by climate change and are involved in setting the agenda are just interested in securing research grants from governments and conspiring to control the people by creating hysteria and fear.

Strangely, the issues of science are being decided based on what most people believe and not by scientific methods. Most of the literature centres around who thinks what, and the literature written by disciples of any discipline other than science in general or climate science, in particular, tends to reflect this. Even the proponents of consensus arguments themselves have no scientific background.

"I have been dismayed over the bogus science and media hype associated with the dangerous human-caused global warming hypothesis. My innate sense of how the atmosphere and ocean function does not allow me to accept these scenarios. Observations and theory do not support these ideas." Professor Emeritus William Grey, Colorado State University, 2006

According to O'Neill and Nicholson-Cole (2009), fear-based messaging about climate change is frequently used in public discourse. While it can capture people's attention, it is often inadequate at eliciting genuine concern for the issue and may even be counterproductive in certain cases. To encourage action, some advocates have turned to fear-appeal frameworks like the Extended Parallel Process Model (EPPM) in climate change communication, emphasising the need to accompany fear-based messages with actionable solutions. The EPPM model lets us know how individuals respond to fear and how much fear has to be applied to elicit the desired outcome. However, as Alison McQueen (2021) pointed out, relying solely on fear-based messaging is incompatible with democratic values and civil discourse.

The message here is to encourage people to educate themselves about climate change before making any bold claims or engaging in scaremongering. It is important to undertake research and gather facts rather than simply blame the human race for climate change without understanding the complexities of the issue. If you are not capable of conducting the research yourself, it is best to question who stands to benefit from the rhetoric and who has an agenda to push. Often, corporations and other entities use crises such as the COVID-19 pandemic as opportunities to create false narratives to further their interests. Therefore, it is important to be aware of unsubstantiated claims and to seek out reliable sources of information to form your opinions and take action.

The argument being made here is that while humans are responsible for air, water, and soil pollution, the claim that human activity is the primary cause of climate change is baseless. This claim is either the result of ignorance or a deliberate ploy by some corporations and politicians to instil fear in ordinary citizens and exploit them for their financial gain through unethical and scandalous business practices.

Scientists who don't rely on research funds to fund their work are concerned about the current political atmosphere due to the general failure to recognise and understand crucial issues. Climate change has become a political tool for labelling sceptics and deniers in an effort to marginalise them and whip up support for the cause. There has been no convincing evidence that any widely promoted global warming models are accurate.

It's important to stop the political hype, hysteria, and fear-mongering driving the climate change agenda and look at all sides of the climate discussion. The common confusion between "weather" and "climate" is one of the first and most crucial truths we must learn. The climate is the statistical average of the weather over a lengthy period of time, often 30 years, while the weather is the short-term variation you can sense when you step outside (Lear, Caroline H. et al., 2020).

In order to gain a comprehensive understanding of the Earth's spheres and their influence on climate, it is imperative to transcend the political hype, hysteria, and fear-mongering surrounding the climate change agenda and recognize the fundamental distinction between "weather" and "climate." The atmosphere, hydrosphere, lithosphere, and biosphere together form what is known as the Earth system or the Earth's spheres.

The atmosphere is the layer of gases surrounding the Earth. It interacts with the other components through processes such as weather patterns, climate regulation, and the exchange of gases with the biosphere. The atmosphere influences temperature, pressure, and weather conditions, providing gases like oxygen and carbon dioxide necessary for life.

The hydrosphere encompasses all the water on Earth, including oceans, lakes, rivers, groundwater, and glaciers. It interacts with the atmosphere through evaporation, condensation, and precipitation, which drives the water cycle. The hydrosphere is essential for supporting life, regulating climate, and shaping the Earth's surface through processes like erosion and sedimentation.

The lithosphere refers to the solid outer layer of the earth, which includes the crust and part of the upper mantle. It interacts with the other components through processes such as the cycling of nutrients, the formation of landforms, and the support of ecosystems. The lithosphere provides habitats for organisms, stores minerals and resources, and undergoes geological processes like plate tectonics.

The biosphere comprises all living organisms, including plants, animals, and microorganisms. It interacts with the other spheres through complex ecological processes. The biosphere depends on the atmosphere for gases, the hydrosphere for water, and the lithosphere for nutrients and habitats. At the same time, the biosphere plays a crucial role in regulating biogeochemical cycles, influencing the climate, and shaping the physical environment.

The interactions and dynamics among the atmosphere, hydrosphere, lithosphere, and biosphere play a significant role in shaping and influencing the climate of the Earth. These components are integral to the Earth's climate system, and changes or disruptions in any of them can have profound effects on climate patterns.

The IPCC, a branch of the United Nations, is considered the cornerstone of discussions on global warming, and its reports are considered gospel and authoritative due to their extensive promotion to the general public. Because it is a political institution with political goals, that is precisely where the difficulty resides. In no way, shape, or form can it be considered scientific? Since it is a political body, its policy advice will necessarily be coloured by partisan considerations.

It is crucial for individuals without specialised knowledge in climate science to seek out information from sources beyond the IPCC framework and reports. The IPCC is an intergovernmental body of the United Nations with representatives from numerous countries worldwide. Since the United States is a significant stakeholder in the United Nations, hosting its headquarters and providing the largest financial contribution to its budget. These official bodies' information can occasionally become distorted by powerful organisations with their own agendas and potential conspiracy theories. Therefore, it is crucial to actively seek out non-governmental, non-US, and non-UN sources of information regarding climate change, thereby avoiding potential bias by obtaining a more balanced perspective.

1.1 The Conspiracies

Conspiracies are plans or actions developed by powerful individuals or groups aimed at gaining financial or strategic advantages while diverting public attention away from relevant issues. Conspiracy theories are the explanations put forward to uncover such motives. Being able to detect a conspiracy requires reading between the lines. While conspiracy theories can play an important role in alerting the public to impending dangers and the hidden motives of powerful groups, it's worth noting that these same groups can sometimes float conspiracy theories as part of their strategy to mislead the public and draw attention away from their activities.

Creating hype around an issue and manipulating the public through media messages to achieve a desired outcome is not a difficult task. In today's world, where a small group of individuals own a substantial portion of the world's wealth through a web of connected shell companies and tax havens, it is essential to be cautious, as they also largely control the media.

One should note that the US and other Western nations have been the primary drivers of heightened attention to climate change, with little regard for the perspectives of the rest of the world. It is well known that a small number of multinational corporations with significant influence and power indirectly rule the US. These same corporations are often accused of promoting conspiracy theories, particularly in relation to the role of fossil fuels in climate change. Interestingly, these corporations stand to benefit regardless of the outcome of the debate, as they are heavily involved in both traditional fossil fuel energy and the newer "clean" energy projects, which receive billions of dollars in government subsidies. It is therefore important to carefully examine who stands to benefit from various climate change policies and initiatives.

A common response to those who question the mainstream narrative on climate change is to label them as sceptics or deniers. However, it is important to recognise that there may be other factors at play, including financial interests and political agendas. It is not unreasonable to ask who stands to benefit from the push for renewable energy and the demonization of fossil fuels. It is important to note that whatever the consequences of climate change, the policies are aimed at helping corporations. While some may dismiss such questioning as conspiratorial, it is important to approach these issues with a critical eye and an open mind.

Conspiracy theories, according to John Cook, a leading proponent of consensus theory, are harmful to society because they are founded on faulty thinking patterns that result from the use of unreliable tools (Lewandowsky, S., and J. Cook, 2022). Furthermore, he insists that conspiracy theories can be used rhetorically, like escapism, to avoid unfavourable conclusions. I don't understand why the consensus brigade keeps pedalling to get their agenda passed if they're convinced their research has already made its point. Does it make a difference whether or not the readers of Aman Tyagi's research paper on climate change are believers or unbelievers (Tyagi, Aman, and Kathleen M. Carley, 2021)? Do we use our faith to decide scientific questions? Have we gone off the rails?

> *"Global warming became a political cause because there was no other enemy following the end of the Cold War."* William M. Gray (1929–2016) was an American atmospheric scientist and professor at Colorado State University.

Conspiracy theorists make two main assertions about climate change: First, the facts about climate change are predetermined to promote ideological purity over scientific rigour. Inhofe has also asked if man-made global warming is "the greatest hoax ever perpetrated on the American people" (Inhofe, J.M., 2012). Martin Durki n, the director of the TV documentary The Great Global Warming Swindle, has described climate change as a "multi-billion dollar worldwide industry" that anti-industrial environmentalists have created. He has suggested that the purpose of climate change is to exercise political influence, introduce world government, and control people.

The notion that climate change is a hoax perpetrated for political gain has been further fueled by the media's portrayal of figures like Al Gore as driving forces behind the global climate movement, thus contributing to the polarisation of opinions on the matter.

Climate change is a complex and multifaceted issue that is perceived, presented, studied, and viewed differently by various interest groups. Conspiracy theories aside, let us look at how these groups perceive, view and present it in the following chapters.

1.2 An Inconvenient Truth

A range of basic human emotions exists, including fear, anger, disgust, happiness, sadness, surprise, and contempt, with fear being the most prominent. Those who set the agenda have been exploiting this emotion of fear in order to elicit a particular response, especially in response to Al Gore's documentary, "An Inconvenient Truth", released in 2006. According to available information, "An Inconvenient Truth" was widely screened and garnered significant viewership. The documentary was screened at various film festivals, educational institutions, community events, and commercial theatres worldwide. While I don't have specific data on the total number of screenings or viewership, the film received substantial attention and was viewed by millions globally. The documentary's impact can also be observed in its box office success. "An Inconvenient Truth" grossed over $49 million worldwide, making it one of the highest-grossing documentaries at the time of its release. The film's popularity also led to its wider availability through DVD releases, television broadcasts, and online streaming platforms, further increasing its viewership.

In many reports on global warming, Al Gore is given prominent placement. This was especially true when climate change was in the spotlight and opinions on the issue among Americans were sharply divided. Al Gore shot to fame amid the chaos surrounding the disputed outcome of the 2000 US presidential election between him and George W. Bush.

The coverage of Al Gore by traditional media was extensive, with Fox News featuring him in 48% of their climate change stories in 2006 and 57% in 2007. Additionally, 28% of the stories in 2006 and 17% in 2007 had explicit references to his movie. Conversely, leading Republican climate change deniers like Senator Jim Inhofe were not featured in any stories on Fox News in 2006, and only 1% in 2007. Gore received several awards and honours, including the Nobel Peace Prize in 2007. In essence, anyone following news about climate change during this time would have been exposed to Gore's message, which was unapologetically pro-climate and called for strong climate action. However, this likely contributed to turning Republicans against the message since they simply saw Gore as a Democratic politician they disliked.

It's implausible that the release of Al Gore's sequel to An Inconvenient Truth, "An Inconvenient Sequel: Truth to Power", released in 2017, will have an impact similar to the original. The movie generated significantly less traction at the box office and in the media. Furthermore, climate change has already become one of the most polarising issues of the day.

There is no way to go back in time. However, this should act as a cautionary tale for the future. It's crucial to select not only a compelling and informative message but also an articulate and persuasive communicator to convey it. The source of Al Gore's credibility as an expert on this topic is not his education or academic credentials but rather his political power, political connections, and vast personal wealth, as there is a lack of any citation in support of him being a climate change expert.

The contrasting approaches of "An Inconvenient Truth" and "The Great Global Warming Swindle" discussed in the following pages serve as a reminder that the messenger's credibility, as well as the content of the message itself, play a crucial role in shaping public perceptions of complex scientific issues like climate change.

1.3 The Great Global Warming Swindle

In 2007, Martin Durkin produced a documentary named "The Great Global Warming Swindle" as a follow-up to Al Gore's film "An Inconvenient Truth." Refuting the supposed scientific consensus on climate change, the movie claims that political and financial interests influence climatology to justify its stance. Despite having a catchy and sensationalist name, the Great Global Warming Swindle is grounded in solid research and interviews with genuine climate scientists. On the other hand, An Inconvenient Truth is primarily an emotional presentation from a single politician.

The documentary showed a variety of individuals, including scientists, economists, politicians, and writers, who rejected the scientific consensus regarding anthropogenic global warming and asserted that man-made global warming is a deception and the greatest con job of our generation. The documentary "The Great Global Warming Swindle" meets Al Gore's "An Inconvenient Truth," which was broadcast on British television and has since been viewed by millions on the Internet.

Here is a quick rundown of the documentary "The Great Global Warming Swindle's" scientific justifications. The evidence does not support the idea that human-caused increases in greenhouse gases are responsible for the recent warming trend. Ice core records have demonstrated that the warming of the oceans is a significant contributor to the rise in atmospheric carbon dioxide, as temperature increases have occurred hundreds of years prior to carbon dioxide increases. These pieces of evidence have been preserved in ice cores and date back over the past 650,000 years. (Fischer, H., et. al. 1999)

Water vapour is by far the most important of the greenhouse gases since it is the most abundant (Ding, J., Chen, J., & Tang, W. 2022). While this paper does not directly state that water vapour is the most important greenhouse gas, it provides relevant information about the relationship between water vapour and carbon dioxide in the atmosphere. Computer climate models are almost exclusively used as the foundation for pessimistic forecasts of future warming. These models, however, do not have a clear understanding of the role that water vapour plays. In any case, we do not have any influence over the water vapour in the air. In addition, computer models are unable to account for the observed cooling that occurred during a significant portion of the 20th century (1940–75) or for the observed warming patterns, which are referred to as the "fingerprints." For instance, the Antarctic is cooling while models predict warming. Yet, while the models predict that the intermediate atmosphere will warm up more quickly than the surface, evidence indicates the complete opposite to be true.

The strongest support for natural causes of temperature changes comes from changes in cloudiness, which closely match regular variations in solar activity. These shifts have been seen for millions of years. Solar activity and the temperature of the Earth are two phenomena that are unique from one another but connected to one another. The so-called Maunder Minimum was a period of very low solar activity that occurred between 1645 and 1715. This period of low solar activity coincided with the coldest excursion of the "Little Ice Age." (Eddy, J. A., 1983) A comprehensive study of long-term solar activity occurred at the same time as the Maunder Minimum. The Zurich sunspot number, a measure of the number of sunspots detected on the sun's surface over a span of eleven years, has demonstrated a noteworthy resemblance between the average temperature of the world's oceans and the extended fluctuations in solar energy (Reid, G. C., 1987). The observation of a striking similarity between these two variables made this discovery possible. The Zurich Sunspot Number is an essential instrument for comprehending the behaviour of the sun and determining the extent of the sun's impact on the climate of the Earth. Scientists can gain a better understanding of the factors that drive the behaviour of the sun and how it may affect Earth's climate if they monitor changes in solar activity over an extended period.

"Solar activity" refers to variations in the total amount of energy that the sun emits over time. This energy is manifested as solar radiation, which encompasses visible light and ultraviolet radiation in addition to various other types of electromagnetic radiation. The activity of the sun fluctuates over the course of an 11-year cycle, with periods of high activity, also known as the solar maximum and times of low activity. The solar maximum occurs once every 11 years.

Friis-Christensen (Friis-Christensen, E., 1993) suggests that the Gleissberg cycle has an influence on the intensity of the eleven-year cycle. The Gleissberg cycle refers to a prolonged solar activity cycle that has a duration of about 80–90 years. It was named after Wolfgang Gleissberg, the first person to describe it. Variations in the number of sunspots seen on the surface of the sun throughout this period are one of the characteristics that define this phenomenon. The 11-year solar cycle is more well-known than the Gleissberg cycle, which is a longer-term pattern overlaid atop the solar cycle. The number of sunspots observed on the surface of the sun swings between relatively low and high levels during the Gleissberg cycle, with the amplitude of this variation being smaller than that of the shorter 11-year solar cycle. The Gleissberg cycle lasts for about 125 million years. Gustav Sporer, a German astronomer, discovered the Gleissberg cycle for the first time at the beginning of the 20th century.

There is some evidence to suggest that the Gleissberg cycle may have an effect on the temperature of the Earth; however, the mechanisms that are responsible for this effect are not completely understood. It has been hypothesised in a number of studies that lower levels of solar activity during the Gleissberg cycle may be linked to lower global temperatures, while higher levels of solar activity may be linked to higher levels of global warming. However, it is essential to keep in mind that the Gleissberg cycle is but one of the many factors that have the potential to affect the climate of the Earth. The Gleissberg cycle is a fascinating facet of solar activity that enables us to gain a deeper comprehension of the behaviour of the sun and how it may have an effect on the climate of the Earth. Alterations in the sun's magnetic field can also have an effect on the amount of cosmic rays that reach Earth, which in turn can have an effect on the development of clouds and the distribution of precipitation patterns.

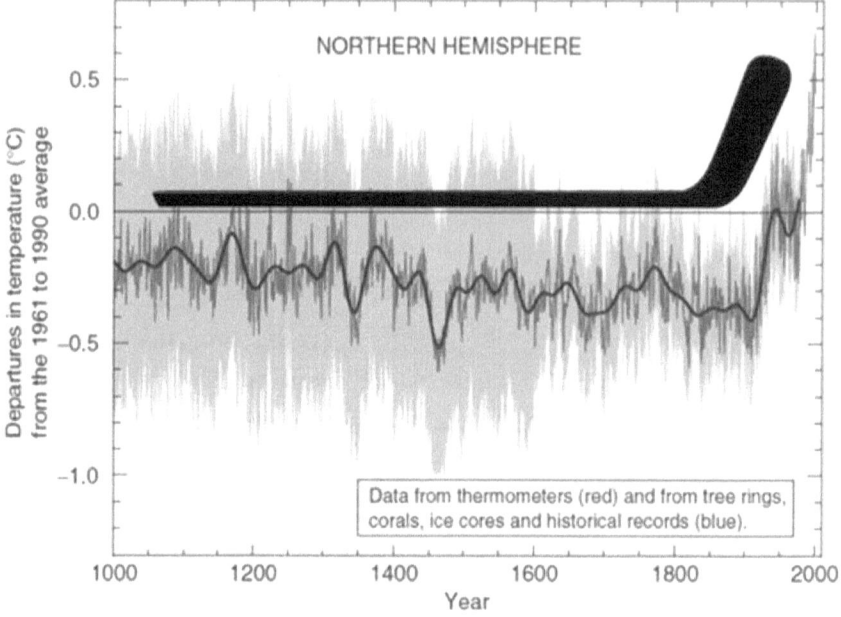

Figure 1: Hockey stick graph

Source: https://bluemarbleearth.wordpress.com/2015/03/11/the-hockey-stick-in-context/

The current warming is most likely part of a natural cycle of warming and cooling that can be traced back approximately a million years. This cycle of climatic warming and cooling can be found all over the world. It explains the Mediaeval Warm Period that occurred about the year 1100 A.D., when the Vikings settled in Greenland during Viking Age (793-1066 BC) and began cultivating the land, and the Little Ice Age, which lasted from about 1400 to 1850 A.D. and was characterised by harsh winters and cold summers across Europe, leading to failed harvests, starvation, disease, and general unhappiness. Opponents have argued that these historically significant climate variations are not real. They rely on a faulty analysis of tree rings and other proxy data, which ultimately resulted in the development of the well-known "hockey-stick" temperature graph, to support their claim that the current warming is "exceptional". At this point, the hockey stick graph has been proven false beyond all reasonable doubt (Montford, A. W. 2010).

The so-called "hockey stick" temperature graph is a graphical representation of changes in the average global temperature over the course of the last thousand years. Because of its shape, which resembles a hockey stick when laid on its side, this type of graph is referred to as a "hockey stick." The chart shows a prolonged period of relatively steady temperatures, followed by a significant increase in temperatures over the course of the last century, which resulted in the upward slope of the "hockey stick" blade. From its initial presentation in a 1998 study by Michael Mann, Raymond Bradley, and Malcolm Hughes, the graph has evolved into a significant piece of data that supports the scientific community's general agreement regarding climate change. The "hockey stick" graph has received much attention and criticism, but it has also received validation and confirmation from later studies that have used a variety of data sources and statistical techniques.

If the majority of the warming is due to natural causes, then there is not much that can be done to stop it. We are powerless in the face of the variable sun, which is thought to be the primary driver of climate change. None of the plans for reducing greenhouse gas emissions, such as controlling carbon dioxide emissions through taxation, rationing, or elaborate cap and trade schemes; promoting uneconomic "alternative" energy, such as ethanol and the impractical "hydrogen economy;" installing massive installations of wind turbines and solar collectors, or proposing projects for the sequestration of carbon dioxide from smokestacks or even from the atmosphere, will be effective because they are all irrelevant, completely ineffective, and incredibly expensive.

In conclusion, no one has been able to demonstrate that a warmer temperature will result in overall negative outcomes. Since the end of the last ice age, which occurred 10,000 years ago, the rate at which the sea level has been steadily increasing has been consistent. This suggests that the much-feared rise in sea levels is not dependent on short-term temperature changes. In fact, a large number of economists contend that the converse is more likely: that global warming results in a net gain, that it raises incomes and standards of living, and that this is the case. Why do we operate under the assumption that the current weather conditions are ideal? The likelihood of this occurring is, without a doubt, infinitesimally remote, and the financial record of previous global warmings corroborates this assertion.

Yet the central idea presented in The Great Global Warming Swindle is far more comprehensive. Why should we spend our limited resources on something essentially a non-problem and disregard the genuine problems that the world is currently facing, like starvation, sickness, and the violation of human rights, not to mention the threats of terrorism and nuclear war? And how well prepared are we to deal with natural disasters, pandemics that can potentially kill most of the human species or even the impact of an asteroid like the one that killed off the dinosaurs? But politicians and elites in most parts of the world would rather waste the limited resources we have on topics that are currently trendy than concentrate on the issues that really need to be addressed. Consider, for example, the terrifying forecasts that are supposedly coming from authoritative figures around the world. The chief scientist of Great Britain warns us that unless we insulate our homes and use light bulbs with a higher efficiency rating, the Antarctic will be the only habitable continent by the year 2100, with only a few surviving breeding couples responsible for the continuation of the human race. Seriously!

The critically acclaimed documentary film An Inconvenient Truth, directed by Al Gore, now has a sequel to Bonni Cohen and Jon Shenk's documentary film "An Inconvenient Sequel: Truth to Power" received somewhat less praise than the first film. This should not come as a surprise, considering that the original film was one of the most financially successful documentaries of all time and dominated the box office on a global scale. In the end, the movie was a contributing factor in Al Gore being awarded the Nobel Peace Prize in 2007 for his work urging action against climate change.

In addition to the many awards and honours it garnered, there is no denying that the movie increased public awareness of the issue of climate change. The documentary definitely had a significantly greater impact on public opinion and knowledge of global climate change than any scientific article or report. In-depth research has been carried out on the manner in which the media handled the topic of climate change during the 1980s, as well as how this coverage may have contributed to the polarisation of public opinion in the United States. The trend that is generally noticed is that public opinion tends to follow rather than lead debate among political elites. This is a common observation. The implications of this finding for our work are significant.

The discussion around climate change has extended beyond politics and academia and has given rise to a range of groups, including the Global Warming Policy Forum, discussed in the following pages, which advocates for a more cautious approach to climate policy and is accused of receiving funding from vested interests.

1.4 The Global Warming Policy Foundation

The Global Warming Policy Forum (GWPF) is a think tank based in the United Kingdom that focuses on energy policy and climate change. Lord Nigel Lawson, who had previously served as the United Kingdom's Chancellor of the Exchequer, established it in 2009. The Global Warming Policy Foundation believes that the widespread scientific consensus on climate change is inaccurate and that the risks connected with global warming have been greatly exaggerated. It warns that policies targeted at decreasing greenhouse gas emissions could have detrimental economic and social repercussions and recommends a more cautious approach to climate policy. Allegations have been made that the Global Warming Policy Foundation gets funding from fossil fuel companies and other organisations with a vested interest in postponing action on climate change. The organisation, on the other hand, has refuted these charges.

The Global Warming Policy Foundation participates in a variety of activities, one of which is the commissioning and publication of research papers, reports, and books on climate change and other related themes. While some of these papers challenge the consensus that the scientific community has reached regarding climate change, others call for a more reasonable approach to solving the issue. The organisation also plans conferences, events, and lectures on the subject of climate change. These events and lectures frequently feature speakers who are critical of the predominant school of thought in climate science or who argue for climate policies that are less stringent. In addition, the GWPF criticises the activities and policies of mainstream climate organisations, such as the Intergovernmental Panel on Climate Change, as well as those of other organisations that advocate for substantial action to be taken on the issue of climate change. Notable publications of the Global Warming Policy Foundation include "The State of the Climate 2019," "The Climate Noose: Business, Nett Zero, and the IPCC's Dangerous Road to Paris," and "The Really Inconvenient Truth or 'It Ain't Necessarily So.'" The organisation also publishes State of the Climate Reports on a regular basis, which are intended to present an alternative view of climate science and policy to the mainstream scientific community. The Global Warming Policy Forum has put together these studies with the help of a group of scientists and academics.

1.5 The Climate Depot

Marc Morano is a conservative political operative who formerly served in the office of Senator James Inhofe (R-Oklahoma). Morano is the brain behind the Climate Depot website. The website presents a number of theories that dispute the scientific consensus on climate change and offer alternative explanations for the phenomenon. On the website Climate Depot, authors who disagree with the findings of mainstream climate science publish articles and pieces of opinion. Climate change researchers and green activists have both criticised it for what they see as its promotion of climate denial and dissemination of false information.

Climate Depot publishes and participates in a variety of activities and publications, including a website with articles and blogs that challenge the accepted scientific understanding of climate change and advocate for denial. This site takes issue with the IPCC's scientific consensus that humans are to blame for climate change. Instead, it asserts that climatic shifts occur as part of a natural cycle. Another illustration is the 2016 release of the Climate Depot film "Climate Hustle." The movie challenges conventional wisdom regarding climate science and defends policies intended to lessen the effects of global warming.

Marc Morano is also a frequent commentator and speaker on climate change. As a means of disseminating his doubts about the consequences of climate change, he has appeared as a guest on numerous radio and television programmes. He has also written several books and articles, including "The Politically Incorrect Guide to Climate Change." This book defends a free-market approach to the issue of climate change and argues that the scientific consensus is politically motivated and unsupported by evidence. According to "Climate Hustle: Are We Being Conned", climate change is a natural and cyclical phenomenon, not a crisis, and the media and political elites are exaggerating the risks of climate change to promote their own agendas. This claim is made within the framework of the theory that climate change occurs in predictable cycles. The final piece, "Green Fraud: Why the Green New Deal Is Even Worse Than You Think", is a rebuttal of the Green New Deal proposal. It argues that the proposal will not be effective in combating climate change and will have disastrous economic consequences.

Besides Climate Depot, many more groups and organisations, like the Global Warming Petition Project, discussed in the following paragraphs, dispute the theory of human-caused climate change.

1.6 The Global Warming Petition Project

Dr Arthur Robinson, a chemist and conservative political activist, launched the Global Warming Petition Project in 1998, also known as the Oregon Petition. It has collected signatures from over 31,000 American scientists who disagree with the theory that human activity is the primary driver of climate change, including 9,029 scientists with PhDs.

In 1998, a petition demanding that the global warming agreement reached in Kyoto, Japan, as well as any other proposals of a comparable nature, be rejected was presented to the government of the United States of America. The signatories to the document argued that the proposed limit on greenhouse gases would be detrimental to the environment, slow down the progression of science and technology, and be detrimental to the health and welfare of mankind. They maintained that there is no compelling scientific evidence to suggest that human emissions of carbon dioxide, methane, or other greenhouse gases are causing or will cause catastrophic heating of the Earth's atmosphere and disruption of the Earth's climate in the foreseeable future. They were referring to the fact that these emissions have already occurred.

The petition aims to demonstrate that the claim of settled science and an overwhelming consensus in favour of the hypothesis of human-caused global warming and the resulting damage to the climate is not accurate. The petition has helped to spark a discussion that is more open and robust than ever before about climate change and the scientific consensus. Additionally, it has inspired additional scientific investigation into climate change and the possible effects that human activities may have on the natural world.

To argue against policy proposals that aim to reduce greenhouse gas emissions, climate change sceptics and policymakers opposed to taking action on climate change have used the petition. The petition's proponents have argued that this has helped to ensure that political agendas are not the only factors influencing public policy decisions but rather scientific evidence as well.

While climate change sceptics have used the Global Warming Petition Project to argue against policy proposals aimed at reducing greenhouse gas emissions, organisations like the Climate and Development Knowledge Network (CDKN), discussed in the following paragraphs, are working towards assisting developing nations in their efforts to combat climate change and advance sustainable development.

1.7 The Climate and Development Knowledge Network

The Climate and Development Knowledge Network, also known as CDKN, is an international organisation that was founded in 2010 with the mission of assisting developing nations in their efforts to combat climate change and advance sustainable development. The governments of the United Kingdom, the Netherlands, and Germany provide funding for the organisation, which offers technical assistance, training, and knowledge-sharing services to governments, non-governmental organisations (NGOs), and other stakeholders in developing countries. The Climate and Development Knowledge Network (CDKN) focuses on four thematic areas: climate-compatible development, climate finance, climate resilience, and climate mitigation. This is done with the intention of assisting developing countries in developing their capacity to respond to the challenges of climate change, reduce greenhouse gas emissions, and build resilience to the impacts of climate change.

The work that the CDKN is doing on climate-resilient development emphasises how important it is to incorporate climate change into development planning and decision-making in order to increase resilience to the effects of climate change and ensure that gains in development are sustainable over the long term. Additionally, the CDKN has supported efforts to promote low-carbon development pathways in developing countries and conducted research on the difficulties developing countries face when attempting to gain access to climate finance. The organisation has also brought attention to the significance of gender considerations in climate change adaptation and mitigation efforts. These efforts include ensuring the participation of women in decision-making processes and addressing the distinct vulnerabilities and requirements of men and women. Since this organisation receives financial support from governments, it seems to have the privilege of propping up and linking with climate change trivial and irrelevant matters such as gender issues. As a result, its work need not be taken seriously.

The work of the Tyndall Centre for Climate Change Research, discussed in the following paragraphs, is complementary to that of the CDKN, as both organisations share the goal of gaining a better understanding of the challenges posed by climate change and developing solutions to those challenges. The Tyndall Centre's focus on multidisciplinary research and collaboration among stakeholders can contribute to the CDKN's efforts to advance sustainable development and address the impacts of climate change in developing countries.

1.8 The Tyndall Centre for Climate Change Research

The Tyndall Centre for Climate Change Research is a research organisation with its headquarters in the United Kingdom dedicated to gaining an understanding of the challenges posed by climate change and finding solutions to those challenges. It was established in the year 2000 and given the name John Tyndall after the physicist John Tyndall, who discovered the greenhouse effect in the 19th century. The Centre facilitates the participation of researchers, policymakers, and other stakeholders from all over the world in order to carry out research that is multidisciplinary in nature and focuses on a variety of issues pertaining to climate change and its effects.

The research conducted at the Centre is primarily focused on three key topics: adaptation, mitigation, and sustainable transitions. It is the purpose of this organisation's research to produce scientific evidence and policy recommendations that can contribute to the mitigation of greenhouse gas emissions, the adaptation of vulnerable populations to the effects of climate change, and the transition to a low-carbon, sustainable future. The Centre engages in knowledge exchange and outreach activities such as workshops, seminars, and publications to raise awareness of the challenges posed by climate change and support the development of evidence-based policies and practices that can help address these challenges. These activities are intended to raise awareness of the issues and contribute to the development of policies and practices that can help address these challenges.

Research on the effects of climate change on various industries, such as agriculture, water resources, and ecosystems, is one of the Center's primary areas of research, and it has produced some important findings. The risks and vulnerabilities associated with climate change have been brought to light as a result of this research, which has also assisted in informing climate adaptation strategies. Research on the transition to a low-carbon economy, including the development of renewable energy sources and the implementation of energy efficiency measures, has also been carried out by the Centre. In addition, the Centre has made a contribution to the creation of carbon budget frameworks and has carried out research on climate justice. This research has focused on the distribution of costs and benefits associated with climate change mitigation and adaptation measures. The findings of this study have brought to light the importance of addressing the unequal distribution of the impacts of climate change and making certain that vulnerable populations are not abandoned during the transition to a low-carbon economy.

The research conducted by the Centre on the effects of climate change on various industries and the transition to a low-carbon economy has significantly contributed to the Global Carbon Project's efforts, discussed in the following pages, to comprehensively understand the global carbon cycle and its impact on climate change.

1.9 The Global Carbon Project

In the year 2001, a group of researchers from all over the world came together to form what is known as the Global Carbon Project (GCP). Its primary purpose is to achieve an all-encompassing comprehension of the global carbon cycle and how it relates to climate change. The Global Carbon Project gathers and analyses data on carbon emissions, atmospheric concentrations, and sinks from all over the world to gain insights into the drivers of global carbon emissions and the impact those drivers have on the climate of the planet.

The GCP conducts research on a wide range of carbon-related issues, including emissions from fossil fuels, deforestation, changes in land use, and the uptake of carbon by oceans and terrestrial ecosystems. The work that it does is geared towards enhancing our understanding of global carbon emissions and the impact they have on climate change.

One of the important deliverables of the GCP is the Global Carbon Budget, which offers yearly estimates of carbon emissions around the world as well as atmospheric concentrations and sinks. These data are utilised by researchers, policymakers, and the media in order to gain an understanding of the trends and patterns of global carbon emissions and to inform policies and actions that are taken to combat climate change.

The Global Carbon Project, in addition to conducting research, offers policymakers and other stakeholders technical support and advice on matters concerning carbon. Its work has contributed to the development of international agreements such as the Paris Agreement, which seeks to keep the increase in the global average temperature below two degrees Celsius. This agreement was one of the outcomes of its work.

The following are noteworthy achievements and contributions attributed to the Global Carbon Project (GCP):

The Global Carbon Project found that global carbon dioxide emissions have increased over the past decade in spite of efforts to reduce greenhouse gas emissions. The amount of carbon dioxide released into the atmosphere reached a new all-time high in 2019, reaching 36.8 gigatons.

According to the Global Carbon Project (GCP), China is the nation that emits the most CO2 into the atmosphere, with the United States, India, and Russia following in that order. These four nations are collectively responsible for more than half of the world's total carbon emissions.

The Global Carbon Project (GCP) has placed a strong emphasis on the significance of changes in land use as a factor in reducing carbon emissions. For example, deforestation is responsible for approximately ten per cent of the world's total carbon emissions.

The Global Carbon Project (GCP) has highlighted the urgent need for significant and rapid reductions in carbon emissions in order to avoid the most severe impacts of climate change. The organisation has advocated for the transition to an economy with lower carbon emissions as well as the implementation of technologies and policies that can cut greenhouse gas emissions.

The Climate Research Unit (CRU) at the University of East Anglia has conducted research providing valuable insights into the factors driving climate change. This reinforces the Global Climate Partnership's call for urgent reductions in carbon emissions and the implementation of effective climate policies and technologies, as discussed in the following pages.

1.10 The Climate Research Unit

The Climate Research Unit (CRU) at the University of East Anglia is widely recognised as a preeminent research unit that focuses on climate science. Since its inception in 1972, it has been conducting research on a wide range of climate-related topics, including the analysis of data pertaining to temperature and rainfall, climate modelling, and the historical reconstruction of climate data. The work it has done has made a sizable contribution to our understanding of the climate system of the Earth as well as the factors that influence climate change.

The creation of the HadCRUT temperature dataset, a global temperature dataset that combines temperature observations from weather stations, ships, buoys, and other sources, is one of the most significant contributions made by the CRU. The HadCRUT temperature dataset was developed in 1979. This dataset is extensively used in climate research and is frequently cited by the Intergovernmental Panel on Climate Change (IPCC) in their assessment reports to better understand the evolution of global temperature over time.

Additionally, the CRU has contributed scientific data and analysis to the reports that the IPCC compiled over the course of its development. Despite this, the Climate Research Unit (CRU) has been the subject of criticism in the past, most notably in connection with the publication of emails leaked in 2009, which allegedly showed evidence of unethical behaviour in the scientific community.

The CRU's research has shown evidence of global warming, which is most likely the result of human activity, such as burning fossil fuels. In addition, the Climate Research Unit (CRU) has made a contribution to the development of climate models, which are used to project future climate scenarios and gain a better understanding of natural climate variability, such as the El Nino-Southern Oscillation (ENSO) and the North Atlantic Oscillation (NAO). The Climate Research Unit (CRU) has worked with other research organisations from around the world to further our understanding of climate change. As a result, we have gained valuable insights into the factors that contribute to climate change and have been able to contribute to the formulation of climate policy on both the national and international levels.

In contrast, the Heartland Institute, discussed below, is sceptical of conventional wisdom regarding climate change and advocates for free-market policies that put economic growth ahead of concerns about the environment while at the same time opposing policies that aim to reduce emissions of greenhouse gases.

1.11 The Heartland Institute

Conservative American think tank Heartland Institute, formed in 1984 and based in Illinois, is noted for its support of free-market policies and scepticism of climate change. It puts economic growth ahead of environmental protection and has aggressively attempted to undermine international climate agreements like the Paris Agreement. At the same time, the institution is against measures that would kerb the release of glasshouse gases.

The Heartland Institute's major argument against climate change is based on its rejection of the field's established scientific consensus. The institute bases its scepticism on the idea that there is still a lot of mystery around the origins and impacts of climate change. It advocates for climate change denial and the assumption that natural processes are the principal drivers of climate change. As a result, it is opposed to measures that would curb GHG emissions, such as the Paris Agreement and carbon price schemes, on the grounds that they would be too expensive and slow economic development.

The Heartland Institute promotes the deployment of alternative energy sources despite its scepticism about and hostility to measures that aim to reduce emissions. It encourages the research and use of alternative energy sources like nuclear power, natural gas, and renewables. Although it supports renewable energy, it is nonetheless critical of efforts to reduce carbon dioxide emissions. Market-based solutions, such as emissions trading programmes and carbon offsets, are advocated by the institute as a means to cut emissions and boost economic growth.

The Heartland Institute's ideology of prioritising free-market principles and restricting government intrusion is consistent with its position opposing carbon reduction initiatives and favouring market-based solutions. This view is shared by think tanks with similar tenets, such as the Cato Institute. Cato Institute, discussed in the following paragraphs, is a libertarian think tank questions established climate science and warned about the unintended consequences of programmes meant to reduce glasshouse gas emissions.

1.12 The Cato Institute

The Cato Institute is a libertarian think tank with its headquarters in Washington, District of Columbia. It is best known for promoting free-market policies and limiting government intervention in economic and social matters. The organisation, which was established in 1977, is frequently critical of policies that aim to reduce greenhouse gas emissions. They frequently cite the perceived costliness of these policies as well as their potential negative impact on economic growth. In addition, it has voiced its doubts about the predominant school of thought in climate science, arguing that there is still a great deal of mystery surrounding the factors that contribute to and the effects of climate change.

The Cato Institute has been active in efforts to challenge policies such as the Clean Power Plan and the Paris Agreement, both of which are aimed at reducing greenhouse gas emissions while at the same time promoting alternative measures such as geoengineering and adaptation. The organisation has been accused of promoting views not supported by scientific evidence and of receiving funding from the fossil fuel industry, who are opposed to it. Despite this, the Cato Institute has maintained that its position on climate change is founded on a dedication to the principles of free markets and limited government intervention.

In a nutshell, the Cato Institute is an influential think tank that defends free-market economic policies and calls for the government to intervene only to a limited extent. Its position on climate change is frequently one of scepticism towards the predominant school of climate science and criticism of policies that aim to cut emissions of greenhouse gases. The organisation advocates for market-based solutions, such as emissions trading schemes and carbon offsets, as well as alternative measures, such as geoengineering and adaptation, as potential solutions to address the effects of climate change.

Similarly to the Cato Institute, the Institute of Public Affairs (IPA), discussed hereafter, is a conservative think tank that supports free-market policies and limited government intervention while conducting research and advocacy on various economic and social issues.

1.13 The Institute of Public Affairs

Since its founding in 1943, the Institute of Public Affairs (IPA) in Australia has been a highly regarded think tank that has made important contributions to the formation of public policy and the advancement of conservative principles. The Institute of Public Affairs (IPA) has been a strong voice in the political landscape of Australia as a result of its emphasis on campaigning for limited government, free markets, and individual liberty.

The Institute of Public Affairs (IPA) is an organisation that, at its heart, advocates for the ideas of classical liberalism and economic freedom. It has a strong faith in the ability of unrestricted free markets to propel economic expansion, bring about prosperity, and encourage innovation. The Institute of Public Affairs (IPA) is an organisation that conducts extensive research and analysis in order to generate policy recommendations. These suggestions place emphasis on deregulatory policies, lower tax rates, and reduced government intrusion as a means to strengthen economic competitiveness and individual liberty.

Beyond the realm of economic concerns lies the IPA's dedication to the liberty of the person. It also emphasises personal liberties, such as the right to free expression and little government interference in the lives of its residents. The institution maintains a persistent commitment to advocating for the adoption of laws that protect civil freedoms such as freedom of expression, freedom to associate, and freedom of choice. It is of the opinion that the cultivation of social advancement and democratic ideals requires the existence of a rigorous civil society that is free from the undue supervision of the government.

The IPA also places a strong emphasis on the promotion of a government that is more accountable to its citizens and more transparent to the public. The institution has a constant policy of advocating for the streamlining of bureaucracy, the elimination of unnecessary red tape, and the improvement of government accountability. Its purpose is to provide individuals and businesses more autonomy by reducing the number of superfluous restrictions that hinder economic activity and innovation.

In addition to the findings of its research and the suggestions it makes about policy, the IPA participates actively in public dialogue and works to provide a forum for a variety of points of view. It routinely organises events, conferences, and public forums that, to examine and debate important problems, bring together specialists, policymakers, and citizens from various backgrounds. The IPA helps foster a vibrant exchange of ideas and contributes to the formation of well-informed public policy in the process of supporting these conversations.

It has been argued that the influence and agenda of the IPA favour the interests of corporations and help to keep existing imbalances in place. They argue that the institute's stance on low government monitoring can lead to market failures and socioeconomic inequities, and they make this argument in a number of different ways. However, proponents of the IPA say that its emphasis on free markets and limited government is necessary for the promotion of economic growth, individual liberty, and prosperity for all parts of society.

Overall, the Institute of Public Affairs has become a major institution in Australia's policy landscape, giving conservative ideas on limited government, free markets, and individual liberties. The commitment of the Institute of Public Affairs (IPA) to promoting informed conversation and advocating for policy improvements has indisputably affected public discourse in Australia for several decades, despite the fact that its views and recommendations may be susceptible to examination and debate.

CHAPTER 2

The Cult

Hype and cult are two concepts that share a close relationship, as both involve a strong emotional attachment to a particular phenomenon or idea. Hype is often used to describe the excessive excitement or publicity surrounding a product, event, or idea, often resulting in a short-term boost in popularity or attention. A cult, on the other hand, refers to a group of people who share a strong attachment to a particular ideology or belief system, often involving a charismatic leader and a strict set of practices or rituals. In some cases, hype can lead to the formation of a cult-like following where individuals become deeply invested in a particular product or idea, often to the point of irrational behaviour or extreme devotion. At the same time, cults may also create hype around their teachings or practices, seeking to attract new followers through a sense of excitement or exclusivity. Both hype and cults can have significant psychological and social effects, influencing individual behaviour and group dynamics in various ways. While I discussed the hype in the previous chapters, I will discuss the cult around climate change in this chapter.

People often use the word "cult" to describe a wide range of religious, spiritual, and philosophical groups because they all have things in common, like authoritarian leadership, strict control over members, and the use of manipulative or coercive tactics.

It's worth noting that some groups may display certain cult-like characteristics, but that does not necessarily mean that they are a full-blown cult. It's important to take a nuanced approach and gather as much information as possible before making any judgements or conclusions about a group.

There is growing concern about the radicalisation of the climate change movement, which has given rise to a new phenomenon, the climate change cult. The climate change cult is a group of individuals fanatically devoted to the idea that climate change is an imminent threat to humanity and that drastic measures must be taken to combat it. While their passion for the environment is commendable, the cult-like nature of their beliefs and actions is worrying. Here, we explore the dangers of extreme environmentalism and the ways in which it threatens our society.

The first problem with the climate change cult is that it often disregards scientific evidence in favour of ideology. Many of its members promote an alarmist view of climate change that ignores the complexity of the issue. They often exaggerate the risks and overlook the potential benefits of technological solutions. This attitude can lead to poor decision-making and misguided policies that harm the environment and society as a whole.

Another issue with the climate change cult is the way it shuts down debate and demonises dissenting voices. Climate sceptics, who question the extent and causes of climate change, are often vilified as "deniers" and accused of being paid off by big oil or other vested interests. This approach is not only unproductive but also dangerous because it creates an echo chamber where only one side of the argument is heard.

The climate change cult's approach to activism is also problematic. Its members often engage in disruptive and illegal activities, such as blocking roads, chaining themselves to buildings, or vandalising property. These actions are not only illegal but also counterproductive, as they turn public opinion against the movement and undermine its credibility.

Moreover, the climate change cult's obsession with climate change often leads to the neglect of other important environmental issues. For example, they may overlook the negative impacts of renewable energy sources such as wind and solar farms on wildlife. This tunnel vision can hinder progress in other areas of environmental protection.

The climate change cult is a dangerous phenomenon that threatens to undermine the progress of the environmental movement. Its fanaticism and disregard for scientific evidence, the demonization of dissenting voices, disruptive activism, and tunnel vision can lead to misguided policies and a lack of progress on other environmental issues. It is essential that we encourage a more rational and evidence-based approach to climate change activism that fosters debate, promotes constructive dialogue, and seeks common ground to tackle this global challenge.

Cult expert and intervention specialist Rick Ross (Ross, R., 1999) enumerates ten warning signs of a group: absolute authoritarianism without accountability; no tolerance for questions or critical enquiry; no meaningful financial disclosure regarding budget or expenses; unreasonable fear about the outside world, such as impending catastrophe, evil conspiracies, and persecutions; there is no legitimate reason to leave; former followers are always wrong in leaving, whether negative or even evil; former members often relate the same stories of abuse and reflect a similar pattern of grievances; there are records, books, news articles, or television programmes that document the abuses of the group or leader; followers feel they can never be "good enough"; the group or leader is always right; and the group leader asserts exclusive authority in determining "truth" or providing validation. This applies perfectly to climate change alarmists. However, it's important to note that these warning signs are not a definitive checklist and that there may be variations, permutations, combinations, or additional signs that could indicate a cult. Additionally, these warning signs should not be applied broadly or used to stigmatise individuals or groups with certain beliefs or opinions.

Many people, from politicians to celebrities to young people, are calling for action to be taken on climate change. Climate preachers like Al Gore, who tell the rest of us to cut back on our carbon footprint while they bask in unmatched opulence, often lack formal scientific training.

Vicious smear campaigns are launched against those who leave, withdraw, or even moderately criticise the climate alarmist movement. After witnessing this phenomenon firsthand, Dutch professor Richard Tol broke ties with the Intergovernmental Panel on Climate Change (IPCC) because of its gloomy predictions. Professor Tol, Dr Richard Lindzen of MIT, Dr Nils-Axel Mörner, and countless other former IPCC climate experts have been the targets of smear campaigns from their peers and the media after they voiced scepticism about the climate change movement's alarmist predictions. For the purposes of the climate change movement, "science" refers to any "science" that backs up the claims of the movement, while "denialism" refers to actual scientific research that disproves the claims of the movement.

The climate change cult's core belief is that humans cause global warming and that taking immediate action is necessary to mitigate its effects. Those who adhere to the tenets of the cult of climate change insist that drastic measures must be taken immediately. They promote actions that would reduce the negative impact of human activities on the environment, such as reducing emissions of greenhouse gases and switching to renewable energy sources. Critics of the climate change cult say it employs doomsday predictions and other scare tactics to further its political goals. This is unnecessary alarmism that serves only to incite panic and a sense of hopelessness rather than rational thought and decisive action.

Chapter 3

The History

3.1 Timeline of key milestones

In this chapter, we'll look at how our current understanding of climate change is the result of a confluence of scientific discoveries, environmental legislation, and world events. We will begin with the earliest scientific warnings about climate change and follow them through to the present, including the major international accords that have been reached. We will also consider the progress made towards understanding the problem and the difficulties still to be overcome. We may learn a great deal about the current status of climate change and the efforts that need to be made to lessen its impact on the world and its inhabitants by looking at a chronology of significant events.

The Mediaeval Warm Period (900–1300) is characterised by warm temperatures across Europe as a result of a pronounced North Atlantic Oscillation.

Circa 1350 to 1850, the northern hemisphere experienced a cooling period known as the "Little Ice Age."

In 1709, as the Little Ice Age was drawing to a close, an abnormally cold winter gripped Europe.

Thomas Newcomen developed the first practical steam engine in 1712, which led to the rise of industry and the widespread adoption of coal.

In the year 1800, the global population hit one billion.

In 1820, French mathematician and physicist Jean-Baptiste Joseph Fourier proposed the concept of the greenhouse effect or the idea that the amount of energy entering the planet as sunlight must be equal to the amount of energy leaving the planet via radiation from heated surfaces. The Earth's atmosphere must absorb some of this heat energy so it doesn't escape into space. He hypothesised that the atmosphere functions like a greenhouse, letting in light but trapping heat inside.

In 1850, Eunice Newton Foote did some research that expanded on the Greenhouse concept. Using glass cylinders, Foote showed that the sun's heating effect was stronger in humid air than in dry air. She found the most intense heating coming from a container of carbon dioxide. Her findings corroborated those of Irish scientist John Tyndall, who also determined which gases were most important in heat absorption.

Water vapour and other gases, as demonstrated by Irish physicist John Tyndall in 1861, caused the greenhouse effect. The Tyndall Centre, a leading UK climate research institution, was named in his honour more than a century after his death. Tyndall also investigated the possibility that different gases absorb different amounts of light. His lab tests revealed that coal gas composed of carbon dioxide, methane, and volatile hydrocarbons is particularly good at soaking up the heat. In the end, he showed that carbon dioxide alone could absorb sunlight of varying wavelengths like a sponge.

In a paper he released in 1863, John Tyndall explains that water vapour can act as a greenhouse gas.

Karl Benz introduced the Motorwagen, widely considered the first true automobile, in 1886.

In 1890, two researchers, the Swedish scientist Svante Arrhenius and the American Thomas Chrowder Chamberlin, independently considered the potential consequences of increasing atmospheric carbon dioxide levels. Both researchers were aware of the link between fossil fuel combustion and climate change, but neither imagined it had already begun.

The average surface air temperature rose by about 0.25 degrees Celsius between 1890 and 1940.

In 1896, Svante Arrhenius wondered if the planet would cool if carbon dioxide levels dropped. He considered the possibility that lower global carbon dioxide levels might explain the ice ages if volcanic activity were to decline. According to his findings, the world's average temperature could drop by about five degrees Celsius if carbon dioxide levels were halved. Arrhenius pondered whether or not the converse was correct. He went back to the numbers and tried to figure out what would occur if carbon dioxide levels were doubled. At the time, this prospect seemed improbable, but his findings indicated that average global temperatures would rise by about 5°C. Decades later, cutting-edge climate modelling proved that Arrhenius' estimates weren't too off.

In 1900, another Swede named Knut Angstrom found that carbon dioxide strongly absorbed portions of the infrared spectrum, even at the tiny concentrations found in the atmosphere. Inadvertently, he proved that even a small amount of gas can cause the greenhouse effect.

In 1927, man-made carbon emissions from burning fossil fuels and industry hit the one billion-tonne mark.

In 1930, the world's population had doubled to two billion since 1800. British engineer Guy Stewart Callendar noted that the United States and the North Atlantic region experienced significant warming during the Industrial Revolution. According to Callendar's estimates, if carbon dioxide levels in the atmosphere were to double, the planet would warm by two degrees Celsius. Even in the 1960s, he maintained his position that the Earth was warming due to the greenhouse effect. Despite widespread scepticism, Callendar was credited with bringing attention to the possibility of global warming. This focus helped secure funding for pioneering efforts to track climate change and carbon dioxide levels in the atmosphere.

In 1938, Guy Stewart Callendar demonstrated that global temperatures had increased over the previous century by compiling data from 147 weather stations across the globe. He also provides evidence that carbon dioxide levels rose during this time and concludes that this was the root cause of the temperature rise. The "Calendar effect" is not taken seriously by most meteorologists because it lacks a scientific basis and is considered a pseudoscientific concept. The Calendar effect suggests that weather patterns or extreme events are influenced by specific dates, such as holidays, seasons, or historical events. However, this notion is not supported by rigorous scientific research or empirical evidence. While there may be anecdotal observations or coincidences of weather events occurring on certain dates, it is important to differentiate between correlation and causation. Just because a weather event coincides with a specific date or historical event does not mean there is a direct causal relationship between them. Without robust scientific evidence, it is challenging to establish a genuine link between calendar dates and weather patterns.

There was 0.2 °C of global cooling from 1940 to 1970. The greenhouse effect is losing its scientific appeal. Some climate scientists have warned of an impending ice age.

In 1955, US scientist Gilbert Plass studied the infrared absorption of different gases with a new generation of equipment, including early computers. According to his calculations, temperatures would rise by 3 to 4 degrees Celsius if carbon dioxide levels were doubled.

Many people assumed that the ocean would be able to soak up all the extra carbon dioxide being released into the atmosphere, but in 1957, US oceanographer Roger Revelle and chemist Hans Suess proved that this wasn't the case.

In 1958, Charles David Keeling began taking systematic readings of atmospheric carbon dioxide at Mauna Loa in Hawaii and Antarctica using equipment he had developed himself. The project, which is still ongoing to this day, provided the first definitive evidence that carbon dioxide concentrations were increasing within four years. The Scripps Institution of Oceanography's 1958 installation of a monitoring station atop Hawaii's Mauna Loa Observatory stands out as the most well-known of these studies related to the Keeling Curve.

In 1960, three billion people called this planet home.

In 1965, experts on the US President's Advisory Committee expressed grave concern over the "greenhouse effect."

In 1969, the first international conference on climate change urged governments to prepare for and prevent any adverse climatic effects that human activity might cause.

Varieties of climate-related anxiety spread widely in the 1970s. Some scientists hypothesise that the increased atmospheric pollution is cooling the planet, prompting widespread public concern. Due to an increase in aerosol pollutants after World War II, the Earth cooled slightly between 1940 and 1970. An article, "Another Ice Age?" appeared in Time magazine in 1974, popularising the theory that sun-blocking pollutants could cause the earth to cool.

Stockholm, Sweden, hosted the first UN environment conference in 1972. Chemical pollution, atomic bomb tests, and whaling take up most of the time, with climate change receiving only a passing mention. As a result, the United Nations established the UN Environment Programme (UNEP).

In 1975, there were four billion people in the world. In the title of his scientific paper, American scientist Wallace Broecker puts the term "global warming" into the public domain.

In 1985, the first major international conference on the greenhouse effect was held in Villach, Austria, where it was warned that greenhouse gases would cause a rise in global mean temperature greater than any seen in human history in the first half of the next century. One metre of sea level rise is possible as a result of this. In addition to carbon dioxide, the conference found that gases like methane, ozone, CFCs, and nitrous oxide all contribute to global warming.

In 1987, five billion people were living on Earth. Thanks to the Montreal Protocol, chemicals that deplete the ozone layer are now restricted. Although it was not created with climate change in mind, it has been more successful at reducing emissions of greenhouse gases than the Kyoto Protocol.

In 1988, the Intergovernmental Panel on Climate Change (IPCC) was established to collect and evaluate data related to climate change and offer a scientific perspective on climate change and its political and economic effects. Drought and wildfires swept across the United States in 1988. The public and the media started paying more attention when scientists started sounding the alarm about climate change. Scientists testifying before Congress in Washington, DC, attributed a severe drought in the United States to global warming, garnering attention from around the world. Scientists at a Toronto conference on climate change later called for a 20% reduction in global carbon dioxide emissions by 2005. In order to analyse and report on scientific findings, the United Nations established the Intergovernmental Panel on Climate Change (IPCC).

In a speech to the United Nations in 1989, Margaret Thatcher, then prime minister of the United Kingdom, warned of a dramatic increase in atmospheric carbon dioxide. As a result, future shifts are likely to be deeper and broader than anything we've experienced before. She advocated for a global climate change treaty. Annual fossil fuel and industrial carbon emissions now total six billion metric tonnes. As the concept of global warming became more widely accepted, scientists began to investigate its potential consequences. According to the forecasts, the seas will get warmer, leading to stronger hurricanes and heat waves. Sea levels are expected to rise anywhere from 28 centimetres to 98 centimetres by the year 2100, according to other studies, which could cause flooding in many cities along the east coast of the United States.

The IPCC issued its first assessment report in 1990. Human emissions are adding to the atmosphere's natural complement of greenhouse gases, which is expected to result in warming, and the results show that temperatures have risen by 0.3–0.6 C over the last century.

The warming trend was halted in 1991 after an eruption of Mount Pinatubo in the Philippines released debris into the stratosphere, blocking some of the sun's rays from reaching Earth. For two years, temperatures generally dropped before beginning to rise again.

The United Nations Framework Convention on Climate Change was established in 1992 at the Earth Summit in Rio de Janeiro. Its primary goal is to maintain atmospheric concentrations of greenhouse gases below levels that would result in dangerous human interference with the climate system. The developed nations have pledged to reduce their carbon output to 1990 levels. The Climate Change Convention, signed by 154 countries in Rio, established an initial goal of reducing emissions from industrialised countries to 1990 levels by the year 2000, with further reductions to follow.

Many members of the Alliance of Small Island States have been concerned about the future of their countries since adopting a demand for emission cuts of 20% by 2005. The resulting sea level rise is predicted to be no more than 20 centimetres.

In 1995, the Intergovernmental Panel on Climate Change issued its second assessment report, which concluded that human activity had a detectable effect on Earth's climate. Some are calling this the "final proof" that human activities are causing climate change. At the first full meeting of the Climate Change Convention in Berlin in March, signatories agreed on the Berlin Mandate. The industrialised countries recognised the urgency of completing negotiations on substantial emission reductions by the end of 1997.

The United States finally agreed to legally binding emissions targets and sided with the Intergovernmental Panel on Climate Change (IPCC) against influential sceptic scientists at the second meeting of the Climate Change Convention in 1996. Scientists warn that most industrialised countries will not meet the Rio agreement to stabilise emissions at 1990 levels by the year 2000, as global carbon dioxide emissions have resumed their steep climb after a four-year pause.

The Kyoto Protocol was signed in 1997. While most developed countries have committed to a five per cent reduction in emissions between 2008 and 2012, these pledges vary widely by country. The United States Senate has already stated that it will not ratify the treaty.

The strongest El Nino on record occurred in 1998, a year that was also the warmest due to global warming. The average global temperature was 0.52 degrees Celsius, higher than the average for the period 1961–1990.

In 1999, there were six billion people in the world.

In 2000, scientists at the IPCC reevaluated projected future emissions and issued a dire warning that, in the worst-case scenario, the world could warm by six degrees Celsius within a century. A string of devastating floods around the world reinforces public concerns that global warming is increasing the risk of extreme weather events. However, negotiations in November in The Hague to finalise the "Kyoto rule book" broke down due to disagreements between the EU and the US. The decision-making process was put off until May 2001.

The United States officially exited the Kyoto Protocol in 2002 under President George W. Bush. The warming experienced in the second half of the 20th century is primarily attributable to human-caused emissions of greenhouse gases, according to the IPCC's Third Assessment Report. The final details of the protocol were settled after lengthy discussions in Bonn in July and Marrakech in November. According to experts, the rich nation signatories' agreed cuts in emissions are less than a third of the original Kyoto promise due to loopholes.

Many countries' legislatures approved Kyoto in 2002, including the European Union and Japan. The complex rules of the protocol must first be ratified by the countries responsible for 55 per cent of emissions from industrialised countries. When the United States and Australia pull out of the agreement, it's up to Russia to decide whether to keep or scrap the pact. Meanwhile, the second warmest year on record occurs, and the Larsen B ice sheet in Antarctica begins to crack.

2003 is the third warmest year on record globally, but it was the hottest summer in Europe in at least 500 years, contributing to an estimated 30,000 deaths. After further investigation, scientists determined that global warming at least doubled the likelihood of a heatwave occurrence. This year's estimated $60 billion in damages from extreme weather is a record high. Greenhouse gas emissions have been increasing at an alarming rate since 2003. Kyoto feels the hot and cold winds from Russia.

A consensus on Kyoto was reached in 2004. In May, the same year, Russian President Vladimir Putin said his country would support the protocol. The protocol will come into effect in 2005 following ratification by the Russian parliament on November 18. A study has found a connection between global warming and the heat wave that occurred in 2003. "The Day After Tomorrow", a recent blockbuster film, uses an extreme case of climate change as the basis for its plot.

For the countries that have signed on to the Kyoto Protocol, its provisions will be legally binding as of 2005. While serving as G8 chair and EU president, Tony Blair, prime minister of the United Kingdom, has chosen climate change as a top priority. The Kyoto Protocol entered into force on February 16. While countries without targets, such as the United States and China, agreed to a non-binding dialogue on their future roles in curbing emissions, Kyoto signatories agreed in December to discuss emissions targets for the second compliance period beyond 2012. Despite opposition, Europe implemented its Emissions Trading Scheme. In terms of global average temperature, 2005 ranks second. Scientists have found a connection between the recent record-breaking hurricane season in the United States, the rapid melting of Arctic sea ice, and the permafrost in Siberia. Scientists are sounding the alarm about the impending collapse of the west Antarctic ice sheet at a crucial climate meeting in Exeter, UK.

According to the Stern Review from 2006, if we do nothing about climate change, it could reduce global GDP by as much as 20%. However, taking action would only cost us about 1% of GDP. Annual fossil fuel combustion and industrial carbon emissions amount to eight billion metric tonnes. In the same year as the release of "An Inconvenient Truth," the Intergovernmental Panel on Climate Change (IPCC) released its third report on climate change, warning that catastrophic consequences will likely result from the unprecedented global warming that has occurred since the end of the last ice age. The EPA's refusal to implement regulations on carbon dioxide emissions led to a challenge in the US Supreme Court. NASA and other US government agencies have been accused of trying to silence climate scientists.

Some sceptics in 2007 would argue that the predictions presented by the IPCC and published in media like Gore's film were exaggerated. Future US President Donald Trump was among those who voiced doubts about climate change. By indicating that there is less than a 10% chance that global warming is not caused by human activities, the IPCC's Fourth Assessment Report contradicts Trump's 2012 tweet, which alleged that the Chinese created the concept of global warming to harm US manufacturing competitiveness. Human-caused global warming is probably due to increasing levels of greenhouse gases in the atmosphere. For their efforts to spread knowledge about the effects of human-caused climate change, the International Panel on Climate Change (IPCC) and former US Vice President Al

Gore each received the Nobel Peace Prize. The cost of stabilising greenhouse gases is estimated at $1830 billion in the IPCC's fourth assessment report, and governments are urged to start planning adaptive measures. The report has been accused of being toned down because it doesn't cover every possible outcome. There will be rapid and permanent climate change, the synthesis report warns. 'The Great Global Warming Swindle', a documentary airing on television, claims that climate science is seriously flawed. The US Supreme Court ruled in April that the EPA can indeed regulate carbon dioxide emissions. Since solar activity has been measured to have decreased since the 1980s, this theory that the sun is to blame for rising temperatures can be laid to rest. Government representatives from around the world agreed on a timeline to establish a post-2012 replacement for the Kyoto Protocol at the annual UN climate summit held in December in Bali. After receiving public jeers, the U.S. delegation finally signed the pledge.

The Keeling project, which began in 1958 and continued until 2008, shows that carbon dioxide concentrations increased from 315 parts per million (ppm) in 1958 to 380 ppm in 2008. Barack Obama, the next president of the United States, has promised to work with other nations to address climate change just two months before he takes office. The protocol, which President Bill Clinton signed, called for a 5.2 per cent reduction in greenhouse gas emissions from 2008 to 2012 in 41 countries and the European Union compared to 1990 levels. As a result of climate change endangering its habitat, the polar bear is protected under the US Endangered Species Act. In response, Alaska has threatened legal action. According to the World Conservation Union, global warming threatens the survival of tens of thousands of species. After being elected president, Barack Obama pledged to increase funding for science, particularly in the areas of climate change and energy technology. He chooses Steve Chu, a renewable energy expert and Nobel laureate, to lead the Department of Energy.

In 2009, the "ClimateGate" scandal began after hackers stole thousands of emails from a server at the University of East Anglia's Climatic Research Unit and posted them online. With high hopes for a new global agreement, 192 governments gathered at the UN climate summit in Copenhagen, but they left with only a controversial political declaration called the Copenhagen Accord. At a conference in December, governments, including the United States, will attempt to negotiate a successor to the Kyoto Protocol. Eric Steig and coworkers provide evidence of warming trends in Antarctica. A break in the fragile ice shelf protecting the Wilkins ice sheet hastens the sheet's demise, and the Arctic continues to warm at an alarming rate. In order to keep global temperatures from increasing by two degrees Celsius or more, a recent study suggests that humans should limit their carbon emissions to no more than one trillion metric tonnes. Alaska hosts a gathering of indigenous people from all over the world to discuss a unified response to climate change. The retreating glaciers have prompted Italy and Switzerland to renegotiate their borders.

To help developing nations green their economies and adapt to climate change, the developed world will begin contributing $30 billion over the course of three years, beginning in 2010. Review after review into "ClimateGate" and the Intergovernmental Panel on Climate Change (IPCC) called for greater transparency but exonerated scientists of any wrongdoing. Instead of collapsing as many had feared, the United Nations summit in Mexico ended with a number of agreements.

Scientists concerned about the "ClimateGate" allegations in 2011 have conducted a new analysis of the temperature record, which shows that the land surface of Earth has warmed over the past century. The number of people in the world is now seven billion. According to the available data, the levels of greenhouse gases are increasing at a faster rate than in previous years.

The summer cover of Arctic sea ice in 2012 was the lowest since satellite measurements began in 1979, at an extent of 3.41 million square kilometres.

The daily mean atmospheric carbon dioxide concentration has exceeded 400 parts per million (ppm) for the first time since 1958, according to data from Hawaii's Mauna Loa Observatory. In the first section of the IPCC's fifth assessment report, scientists state with 95% confidence that human activities since the 1950s have been the primary contributor to global warming.

In 2015, the United States signed the Paris Climate Agreement under President Obama. The 197 countries signed the agreement committed to reducing their own greenhouse gas emissions and reporting on their progress. The Paris Climate Agreement hinged on a promise to halt a rise in global average temperatures of more than two degrees Celsius. Many scientists believe a temperature increase of more than two degrees Celsius is dangerous because it increases the likelihood of deadly heat waves, droughts, storms, and rising sea levels. With Trump in the White House, the United States officially withdrew from the Paris Agreement. Trump, the president of NASA and NOAA, both independently confirmed in 2016 that global surface temperatures in that year were the highest since modern record-keeping began in 1880. The United Nations' Intergovernmental Panel on Climate Change issued a report in October 2018 that said "rapid, far-reaching" action was required to keep global warming below 1.5 degrees Celsius and avoid the worst possible, irreversible consequences for the planet.

Teenager Greta Thunberg of Sweden goes on strike over climate change in 2018. By November 2018, over 17,000 students from 24 countries had joined climate strikes in response to her campaign to raise awareness about global warming. Thunberg had already been considered for the Nobel Peace Prize by March 2019. She famously sailed across the Atlantic instead of flying to New York City in August 2019 to attend the United Nations Climate Summit. "You have stolen my dreams and my childhood with your empty words," she told world leaders at the United Nations in September 2019. There is a mass extinction starting right now, and all you can talk about is money and economic growth that never ends. Oh, the gall! It was reaffirmed at the UN Climate Action Summit that "1.5 °C is the socially, economically, politically, and scientifically safe limit to global warming by the end of this century", and a goal of net zero emissions was set for the year 2050.

On his first day in office, January 20, 2021, President Joe Biden signed an executive order to re-enter the agreement. On February 19, 2021, the United States formally re-entered the Paris Agreement.

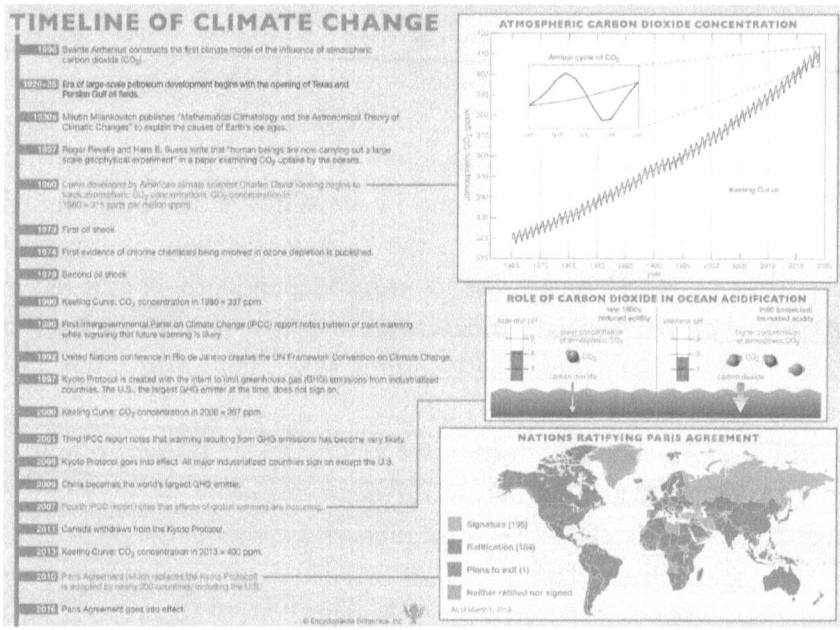

Figure 2: Timeline of Climate Change

Source: https://cdn.britannica.com/92/192592-004-E8A9C1D0/timeline-developments-climate-change.jpg

3.2 History of Climate Change

Climate change refers to the long-term alteration in weather patterns, encompassing variations in temperature, precipitation, and other indicators. This phenomenon has been observed throughout the history of the Earth, where natural factors such as volcanic activity and fluctuations in solar radiation have contributed to changes in the climate over thousands of years. Understanding the history of climate change involves a complex examination of scientific and social factors, leading to our current comprehension of this global issue. Although natural factors have played a significant role in shaping the Earth's climate over millions of years, here is a succinct summary of the progression of climate change.

During the early stages of Earth's existence, its atmosphere primarily consisted of carbon dioxide and minimal amounts of oxygen, resulting in a climate vastly distinct from the present. Moreover, the temperature was considerably higher than it is today. Evidence of climate change in the past can be gleaned from the study of ice cores, which provide records of past temperatures and atmospheric conditions dating back hundreds of thousands of years. These cores reveal that the Earth has undergone numerous cycles of warming and cooling, with the most recent ice age concluding approximately 10,000 years ago. The primary drivers of these long-term climate cycles have been identified as changes in the Earth's orbit and solar radiation.

Over the course of 2.6 million years, the Earth has experienced several ice ages, during which substantial portions of the planet were covered in ice. Variations in the Earth's orbit and tilt, influencing the amount of solar radiation received, were responsible for these cycles of glaciation and deglaciation.

The mediaeval warm period, occurring from around 950 to 1250 A.D., was characterised by relatively warm temperatures, particularly in Europe and North America. This era is also known as the "Mediaeval Warm Period." From the 16th to the 19th centuries, the Northern Hemisphere, in particular, underwent colder temperatures, referred to as the "Little Ice Age."

The advent of the Industrial Revolution in the 18th and 19th centuries marked a significant turning point in human history. The utilisation of fossil fuels and industrialisation led to a substantial increase in the emission of greenhouse gases. In the 20th century, scientific observations began to indicate that human activities were exerting a substantial impact on the Earth's climate. The burning of large quantities of fossil fuels like coal and oil by humans, beginning in the mid-19th century, released carbon dioxide into the atmosphere, thus initiating the contemporary era of climate change.

Since the onset of the Industrial Revolution, the average global temperature has risen by approximately one degree Celsius. An analysis of global temperature data spanning from 1880 to 2016 revealed that the Earth has warmed by roughly 1.1 °C since preindustrial times (Hansen et al., 2016).

3.3 Climate of History

The term "climate of history" is not commonly used in the field of history. However, one interpretation of this phrase might be the idea that the climate and environment in which historical events and processes take place influence and have an impact on them.

For example, environmental factors such as climate change, natural disasters, and resource availability have played a significant role in shaping human history. The rise and fall of ancient civilisations, for instance, can be attributed to changes in climate and the availability of water and other resources. The Little Ice Age, a period of cooling that occurred between the 16th and 19th centuries, had far-reaching effects on Europe, including crop failures, famines, and social unrest.

Conversely, human activities have also had a significant impact on the climate and environment, which in turn have influenced historical events. The concept of the climate of history emphasises the complex and interrelated nature of human societies and the natural world and the ways in which they have shaped each other over time.

Dipesh Chakrabarty (Chakraborty, D., 2009), an eminent historian and postcolonial theorist, has made significant contributions to our understanding of history, climate change, and their interconnectedness. In his influential work, "The Climate of History: Four Theses," Chakrabarty argues that the advent of climate change calls for a reevaluation of historical narratives and a reconceptualization of the relationship between humans and the environment. In this chapter, we will explore Chakrabarty's four scenarios and their implications for our understanding of history and the challenges posed by climate change.

Thesis 1: Nature as a Static Stage

Chakrabarty's first thesis challenges the traditional understanding of history, which often depicts nature as a static backdrop to human events. He argues that this view neglects the significant impact of climate change on historical processes. Chakrabarty asserts that climate change disrupts the stability of the natural environment and, consequently, reconfigures human history. This scenario prompts historians to reconsider the influence of climate on the rise and fall of civilisations, the development of economies, and social transformations.

Historians think history is a natural physical process where nature has no humanistic characteristics like intention and insight. But climate change has changed this equation and highlighted the role of humans as geological agents who have joined nature. This development has removed the distinction between humans, particularly from the historian's point of view.

This merging of natural history and human history is the recognition that human history is deeply intertwined with the natural world and that the study of human history cannot be separated from the study of the natural environment.

In the past, the study of human history and the study of the natural world were often considered separate disciplines, with little recognition of the interconnectedness of the two. However, this view is changing, as scholars are increasingly recognising that humans are part of the natural world and that human history has been shaped by natural processes such as climate change, geology, and biology.

One example of the merging of natural and human history is the study of environmental history, which examines the interactions between human societies and the natural world over time. Environmental historians examine how human societies have shaped the natural environment and how environmental change has influenced human history. This includes topics such as the impact of agriculture on ecosystems, the role of climate change in the rise and fall of civilisations, and the impact of resource extraction on indigenous communities.

Another example is the emerging field of "Anthropocene Studies," which explores the ways in which human activities have influenced the Earth's ecosystems and geological processes, leading to a new geological era characterised by human dominance of the planet.

The merging of natural history and human history is important because it helps us better understand the complexity of human societies and their relationship with the natural environment. It also highlights the need for more sustainable and responsible approaches to human development that take into account the long-term impacts of our actions on the environment and the natural world.

Natural history primarily concerns the study of the natural world, including the Earth's physical and biological phenomena, such as geology, ecology, zoology, botany, and palaeontology. It encompasses the examination of the Earth's natural processes, the diversity of life forms, and the interactions between organisms and their environment. Natural history explores the evolution, behaviour, and distribution of species, as well as the geological formations, climate patterns, and ecosystems that shape the planet.

On the other hand, human history, also known as recorded history or history proper, pertains to the study of human societies, cultures, and civilisations. It encompasses the examination and interpretation of past events, developments, and achievements of human beings. Human history delves into the study of social, political, economic, and cultural aspects of different civilisations, the evolution of human societies over time, and the causes and consequences of historical events and processes.

While natural history focuses on the broader natural world and its non-human inhabitants, human history centres specifically on human experiences, actions, and their impact on society. Both disciplines are crucial for understanding the complexities of our world, with natural history providing insights into the natural environment in which human history unfolds.

Thesis 2: Humans as Universal Actors

Chakrabarty's second thesis challenges the prevailing notion that humans are universal actors independent of their ecological contexts. He highlights the unequal distribution of the impacts of climate change, with marginalised communities suffering the most severe consequences. By acknowledging the differential vulnerabilities and responsibilities in the face of climate change, Chakrabarty calls for a more inclusive and justice-oriented approach to history. This scenario emphasises the need to recognise and amplify the voices of those most affected by climate change.

The human instincts of freedom, reason, and modernity, whereby humans think of and reason out fossil fuels as free from nature; in fact, fossil fuels are extractive in nature. Climate change and globalisation have brought home the point that fossil fuels have a price to pay in the form of destabilisation of the balance and the creation of climate issues.

Modernisation and globalisation are characterised by rapid social, political, and economic change, often associated with the rise of industrialisation and the enlightenment, and the interconnectedness of the economies, societies, and cultures of the world, made possible by developments in communication and technology.

One of the defining features of modernisation is the emergence of nation-states, which have become the dominant political units in the world. Nation-states have facilitated globalisation by creating formal systems of international trade, finance, and diplomacy that allow for the exchange of goods, ideas, and people across borders. This has led to the development of global supply chains, international organisations, and cultural exchanges that have transformed the world's economies and societies.

Globalisation has also been driven by technological advancements, such as the Internet, which have facilitated the exchange of information and communication worldwide. This has led to the creation of global networks and communities that transcend national boundaries and have facilitated the spread of ideas, beliefs, and values across the world.

However, the effects of globalisation have not been evenly distributed, and many argue that it has led to increasing economic inequality and social polarisation. Some have also criticised globalisation for undermining traditional cultures and ways of life and contributing to the degradation of the natural environment.

In summary, modernity and globalisation are deeply intertwined, as globalisation is seen as a product of the rapid social, political, and economic changes that characterise modernity. While globalisation has brought many benefits, it has also contributed to economic inequality, social polarisation, and environmental degradation. As such, it is important to evaluate and manage the effects of globalisation to ensure that they benefit all members of society and the natural world.

Thesis 3: History as the Story of Capitalism

The third thesis of Chakrabarty challenges the widely held belief that economic and human agency has had a major influence on history. He argues that climate change reveals the entanglement of capitalism and the environment. Historically, capitalism's relentless pursuit of growth has contributed to ecological degradation and climate change. This scenario demands a critical examination of the role of capitalism in shaping historical processes and highlights the urgent need for sustainable alternatives.

The climate changes of Anthropocene origin laid bare the side effects of capitalism, its consumeristic pursuits, and the overexploitation of natural resources. The adverse effects of climate change are a direct outcome of these capitalistic activities. Though this is the creation of rich nations, the blame is attributed to the entire human species. Apparently, humanity is highly polarised between rich and poor, the rich global north and the poor global south.

Capitalism and the human species are not necessarily in opposition to one another, but there can be tensions between the two.

On the one hand, capitalism has been instrumental in driving economic growth and innovation and improving the quality of life for many people around the world. It has allowed for the development of new technologies, increased access to goods and services, and improved standards of living. In this sense, capitalism can be seen as a force that has benefited the human species.

However, on the other hand, the pursuit of profit and growth under capitalism has also led to exploitation, inequality, and environmental degradation. The focus on maximising profits and economic growth can come at the expense of the well-being of workers, communities, and the natural world. This can have negative impacts on the human species, such as increased poverty, health issues, and the loss of biodiversity.

Furthermore, the capitalist system is predicated on constant growth, which may not be sustainable in the long term. As resources become scarce and environmental pressures increase, capitalism may be forced to confront its limits and adapt to a more sustainable economic model.

In summary, capitalism and the human species have a complex relationship that is both cooperative and conflicting. While capitalism has brought many benefits to humanity, it also has the potential to harm both people and the environment if it is not managed responsibly. As such, it is important to continuously evaluate and adjust our economic systems to ensure they serve the best interests of all human species and the natural world.

Thesis 4: A New Cosmopolitics

Chakrabarty's fourth thesis proposes a new cosmopolitics that transcends traditional boundaries and recognises the interconnectedness of human and non-human lives. He argues that the challenges posed by climate change necessitate collective action beyond the confines of nation-states. This scenario calls for a global response, where the boundaries between humans, animals, and ecosystems blur, fostering a sense of shared responsibility for the planet. Chakrabarty advocates for cosmopolitics that acknowledges our dependence on the Earth's resources and the need for ecological justice.

The present is the key to history. The study of the history of the human species and the history of capitalism has enabled us to understand the juncture of climate change at which we are at the moment. To understand what the human species has in store for the long term, we need to study this cross-hatching of history, the human species, and the capital in detail.

The cross-hatching of species history and the history of capital refers to the intertwined relationship between the evolution of species and the development of capitalist economies.

On the one hand, the economic systems in which they exist have shaped the evolution of species. For example, the selective pressures of competition for resources in capitalist economies have played a role in the evolution of certain species, such as those that are able to exploit new ecological niches or adapt to changing environmental conditions.

On the other hand, the natural world has also had an impact on the history of capitalism, with the exploitation of natural resources and the destruction of ecosystems having serious consequences for both the environment and the economy. For example, the exploitation of oil and other natural resources has been a major driver of economic growth, but it has also led to environmental degradation and climate change.

The cross-hatching of species history and the history of capital is a reminder that humans and other living beings are deeply interconnected and that the health of our economies and societies is dependent on the health of the natural world. It also highlights the need for more sustainable economic models that take into account the long-term impacts of our actions on both the economy and the environment.

Dipesh Chakrabarty's four scenarios presented in "The Climate of History" compel us to rethink the traditional ways in which we understand and narrate history. By foregrounding the impact of climate change, he calls for a more inclusive, justice-oriented, and sustainable approach to historical scholarship. Chakrabarty's work challenges us to recognise the urgent need for collective action and a cosmopolitical consciousness that transcends national boundaries. Ultimately, his ideas invite us to reimagine history as a dynamic process deeply intertwined with the complex interactions between humans and their environment.

CHAPTER 4

The Politics

This chapter delves into the intricate politics that revolve around the concept of human-caused climate change, exploring the ongoing debates, vested interests, and far-reaching policy implications associated with this understanding. The issue of climate change has transcended scientific realms and become deeply entwined with political ideologies, shaping responses and actions accordingly. Consequently, the politics surrounding human-caused climate change extend beyond national boundaries, influencing international relations as well.

Two prominent institutions, the Intergovernmental Panel on Climate Change (IPCC) and the National Aeronautics and Space Administration (NASA), have emerged as significant players seeking to shape discourse and public opinion due to their substantial influence and political clout. Understanding the extent of their impact and comprehending the nature and scope of their work is crucial to grasping the broader dynamics at play.

The politics surrounding human-caused climate change have transformed the issue into a highly contentious and polarising topic. Diverse ideological perspectives influence the response to the hypothesis of human-caused climate change, with contrasting viewpoints shaping policy decisions and public discourse. Progressive factions tend to prioritise urgent climate action, acknowledging the intergenerational responsibility at hand, while conservative voices may question the scientific consensus and advocate for market-based solutions or limited government intervention.

The international arena is not exempt from the political ramifications of human-caused climate change. Negotiations on climate agreements, such as the Paris Agreement, exemplify the complex power dynamics at play. Developed and developing nations often hold differing positions, with historical emissions as a basis for debates on shared responsibilities, emission reduction targets, and financial and technological support. The politics surrounding climate change significantly impact the ambition, commitments, and enforcement mechanisms encompassed within international agreements.

Furthermore, it is impossible to ignore the significant roles played by organisations like NASA and the IPCC. Their immense size and political might position them as key actors in the climate change discourse. The IPCC, through its comprehensive assessments and scientific research, provides a platform for informed policy discussions and recommendations. With its vast expertise in Earth science and satellite observations, NASA contributes valuable data and analysis to further our understanding of climate change. Both institutions exert substantial influence on shaping public opinion, policy decisions, and the direction of climate action.

By examining the intricate politics surrounding the premise of human-caused climate change, we gain insight into the broader context in which this issue is situated. Understanding the debates, interests, and policy ramifications associated with human-caused climate change is essential for forging effective strategies and collaborative approaches to mitigate its impact. By acknowledging the influence of political dynamics on climate change discourse, we can navigate the complexities of the issue and strive towards collective action for a more sustainable future.

4.1 The Intergovernmental Panel on Climate Change

The Intergovernmental Panel on Climate Change (IPCC) was established in 1988 by the World Meteorological Organisation (WMO) and the United Nations Environment Programme (UNEP) to provide policymakers with regular scientific assessments of climate. The IPCC's role is to assess the scientific, technical, and socio-economic information relevant to understanding the risks associated with human-induced climate change.

The IPCC is made up of three working groups, a task force, a bureau, and a panel. Working Group I focuses on the physical science basis of climate change, including observations of climate change, climate processes, and feedback mechanisms. Working Group II assesses the impacts of climate change on humans and natural systems, including vulnerability and adaptation. Working Group III focuses on mitigation options and strategies for reducing greenhouse gas emissions.

The Task Force on National Greenhouse Gas Inventories oversees the IPCC's Guidelines for National Greenhouse Gas Inventories, which provide guidance to countries on how to estimate and report their greenhouse gas emissions and removals. The Task Force also supports capacity building and training for countries to improve their greenhouse gas inventories.

Greenhouse gas inventories are comprehensive accounts of the emissions of greenhouse gases from a particular source, sector, or country over a given period. These inventories typically include data on the amounts of greenhouse gases released into the atmosphere as a result of human activities such as burning fossil fuels, deforestation, and industrial processes.

Greenhouse gas inventories are used to monitor and track the emissions of greenhouse gases, which are responsible for causing climate change. By quantifying emissions from different sources, greenhouse gas inventories help policymakers, businesses, and individuals understand where emissions come from and identify opportunities to reduce them.

Several internationally recognised protocols and guidelines for developing greenhouse gas inventories include the Intergovernmental Panel on Climate Change (IPCC) Guidelines for National Greenhouse Gas Inventories and the ISO 14064-1 standard for greenhouse gas accounting and verification. These protocols provide standardised methodologies for calculating emissions and reporting them in a consistent and transparent manner.

The Bureau oversees the day-to-day operations of the IPCC, including the selection of authors for IPCC reports, the management of the IPCC's budget, and coordinating the IPCC's activities. The Bureau is made up of the IPCC Chair, the IPCC Vice-Chairs, the Working Group Co-Chairs, and the Task Force Co-Chairs.

The Panel is made up of representatives from 195 member countries, who are responsible for approving IPCC reports and making decisions about the IPCC's future work. The Panel meets every few years to review the IPCC's activities and adopt new work programmes and budgets.

The IPCC operates on a six-year assessment cycle, with each cycle producing a series of reports that cover different aspects of climate change. The reports are divided into three working groups:

- Working Group I: The Physical Science Basis
- Working Group II: Impacts, Adaptation and Vulnerability
- Working Group III: Mitigation of Climate Change

The IPCC Assessment Reports are a series of comprehensive reports produced by the Intergovernmental Panel on Climate Change (IPCC) that provide an up-to-date scientific understanding of climate change. The reports are intended to provide policy-relevant information to decision-makers, governments, and stakeholders to support informed decision-making related to climate change. The IPCC also produces a synthesis report that integrates the findings of the three working group reports and provides a summary for policymakers. The IPCC's reports have been instrumental in shaping international climate policy, including the United Nations Framework Convention on Climate Change (UNFCCC) and the Paris Agreement.

It is worth noting that the IPCC is not a research organisation and does not conduct ongoing research on the subject of the causes of climate change. Instead, the IPCC provides assessments of the current state of scientific knowledge on climate change, including its causes, impacts, and potential mitigation and adaptation measures, based on the best available scientific and technical information from around the world. These assessments are produced by working groups of hundreds of experts in various disciplines and regions who review and synthesise the relevant scientific literature and data.

The staff of the IPCC is made up of a combination of paid professional staff and volunteer scientists from around the world. The paid professional staff at the IPCC Secretariat consists of approximately 70 full-time employees who support the organisation's operations. The volunteer scientists who work for free and donate their time and expertise to the IPCC do most of the work.

The IPCC is funded by contributions from member countries and organisations. The budget for the IPCC varies from year to year, depending on the level of activities planned and the availability of funding. The budget for the IPCC's Sixth Assessment Report cycle, which covers the period from 2015 to 2022, was around USD 195 million. This funding was used to support a range of activities, including the production of assessment reports, capacity building and training, and outreach and communication activities. This budget is used to support the work of the IPCC, including the preparation of assessment reports, technical papers, and other publications, as well as the organisation of meetings and workshops.

It is understood that the IPCC operates on a relatively small budget (195 million) compared to other international organisations.

For example, the budget of the World Health Organisation (WHO) is around USD 5.6 billion, and the budget of the United Nations Development Programme (UNDP) is around USD five billion. However, despite its modest budget, the IPCC has been able to produce some of the most comprehensive and widely respected assessments of climate change science, impacts, mitigation, and adaptation measures.

The IPCC does not conduct its own research; its role is to assess and synthesise the existing scientific literature on climate change. Therefore, the majority of the staff at the IPCC Secretariat are experts in fields such as economics, policy, and communication, as well as administrative and support staff, and as such, the organisation does not have any scientists in its roles.

It is crucial to acknowledge that the IPCC relies solely on scientists who volunteer their time and come from diverse backgrounds and affiliations. Apparently, these scientists depend on grants to fund their research, and it's essential to understand how these grants are awarded. Grant allocation is limited to individuals who support a particular viewpoint, and those with contrasting opinions are ineligible.

The IPCC's review process is designed to be transparent and inclusive, with opportunities for both expert and government review at multiple stages. This helps to ensure that a wide range of perspectives and viewpoints are considered and that the final reports are based on the best available science. It's worth noting that the IPCC relies on volunteer scientists to review their work. It can be difficult to understand how the IPCC's non-scientific team of 70 staff members manages the task of handling, organising, and synthesising such a massive amount of scientific data.

The IPCC has produced five assessment reports to date, with the most recent one released in 2014. Each assessment report is divided into three working group reports and a synthesis report and covers the following areas:

- **Working Group I: The Physical Science Basis** - This report focuses on the physical aspects of climate change, including the changes in the atmosphere, oceans, and cryosphere, and the observations, modelling, and projections of future climate change.
- **Working Group II: Impacts, Adaptation, and Vulnerability** - This report assesses the impacts of climate change on natural and human systems, including ecosystems, water resources, human health, and infrastructure. It also examines options for adaptation and the potential for vulnerable regions and populations.
- **Working Group III: Mitigation of Climate Change** - This report focuses on options for reducing greenhouse gas emissions and limiting the magnitude and impacts of climate change, including technological, social, and economic factors.
- **Synthesis Report** - This report integrates the findings of the three Working Group Reports and provides an overarching summary for policymakers, highlighting the key messages and policy implications of the scientific evidence.

The IPCC Assessment Reports are widely considered to be the most comprehensive and authoritative assessment of climate science available. They have been instrumental in shaping international climate policy, including the United Nations Framework Convention on Climate Change (UNFCCC) and the Paris Agreement. As usual, the Sixth Assessment Report (AR6) was released in three parts. The first part, the Working Group I contribution titled "The Physical Science Basis", was released on August 9, 2021. The second part, the Working Group II contribution titled "Impacts, Adaptation and Vulnerability", was released on February 11, 2022. The third part, the Working Group III contribution titled "Mitigation of Climate Change," is expected to be released in March 2022.

The Intergovernmental Panel on Climate Change (IPCC) and the Conference of the Parties (COP) are two key institutions involved in addressing climate change and formulating global climate policy. The IPCC provides objective and comprehensive scientific assessments on climate change, while the COP is the supreme decision-making body of the United Nations Framework Convention on Climate Change (UNFCCC). The IPCC provides scientific expertise and assessments to inform climate policy discussions and decisions at the COP meetings, while the COP relies on the IPCC's reports and findings to develop effective and evidence-based strategies to address climate change at the global level.

The Conference of Parties (COP) is an annual meeting of the member countries of the United Nations Framework Convention on Climate Change (UNFCCC). It is attended by representatives of member countries, observer states, intergovernmental and non-governmental organisations, and the media. The meetings are organised into various plenary sessions, attended by all parties, and numerous smaller meetings and working groups. The UNFCCC is an international treaty established in 1992 with the objective of stabilising greenhouse gas concentrations in the atmosphere and preventing dangerous human interference with the climate system.

The COP is the supreme decision-making body of the UNFCCC and is responsible for reviewing and guiding the implementation of the Convention. The first COP was held in Berlin in 1995, and since then, the COP has been held annually in different cities around the world.

The main objectives of the COP are to:

- Review the implementation of the Convention and its Kyoto Protocol
- Develop and negotiate new agreements and protocols on climate change
- Establish rules and procedures for implementing the Convention and its protocols
- Provide guidance to the subsidiary bodies of the Convention

Some of the key outcomes of the COP meetings include the adoption of the Paris Agreement in 2015, which sets a long-term goal of limiting global temperature rise to well below 2°C above pre-industrial levels and pursuing efforts to limit it to 1.5°C. The COP also establishes the rules and guidelines for implementing the Paris Agreement and monitors progress towards meeting the agreed-upon targets and goals.

Here is a comprehensive list of the Conference of Parties (COP) to the United Nations Framework Convention on Climate Change (UNFCCC):

1. COP 1: Berlin, Germany (1995)
2. COP 2: Geneva, Switzerland (1996)
3. COP 3: Kyoto, Japan (1997)
4. COP 4: Buenos Aires, Argentina (1998)
5. COP 5: Bonn, Germany (1999)
6. COP 6: The Hague, Netherlands (2000)
7. COP 6 (Part II): Bonn, Germany (2001)
8. COP 7: Marrakech, Morocco (2001)
9. COP 8: New Delhi, India (2002)
10. COP 9: Milan, Italy (2003)
11. COP 10: Buenos Aires, Argentina (2004)
12. COP 11: Montreal, Canada (2005)
13. COP 12: Nairobi, Kenya (2006)
14. COP 13: Bali, Indonesia (2007)
15. COP 14: Poznan, Poland (2008)
16. COP 15: Copenhagen, Denmark (2009)
17. COP 16: Cancun, Mexico (2010)
18. COP 17: Durban, South Africa (2011)
19. COP 18: Doha, Qatar (2012)
20. COP 19: Warsaw, Poland (2013)
21. COP 20: Lima, Peru (2014)
22. COP 21: Paris, France (2015)

23. COP 22: Marrakech, Morocco (2016)
24. COP 23: Bonn, Germany (2017)
25. COP 24: Katowice, Poland (2018)
26. COP 25: Madrid, Spain (2019)
27. COP 26: Glasgow, Scotland (2021)

Each COP has played a crucial role in advancing international action on climate change and setting the stage for future progress. The adoption of the Paris Agreement at COP 21 in 2015 was a major milestone in the global effort to combat climate change, and subsequent COP meetings have focused on implementing and strengthening this agreement.

A Bureau that oversees the IPCC's work upholds its scientific integrity, and facilitates communication with governments and other stakeholders is in charge of running the organisation. The IPCC Plenary, which consists of representatives from all of the IPCC's participating governments, elects the 34 members who make up the Bureau. The Plenary elects a Chair to serve a four-year term as the Bureau's leader.

As of September 2021, the current Bureau of the IPCC consists of the following 34 members:

1. Hoesung Lee (Republic of Korea): IPCC Chair
2. Thelma Krug (Brazil) - Vice-Chair
3. Ko Barrett (United States of America) - Vice-Chair
4. Youba Sokona (Mali) - Vice-Chair
5. Edvin Aldrian (Indonesia)
6. Eduardo Calvo (Chile)
7. Taha Zatari (Morocco)
8. Jean-Pascal van Ypersele (Belgium)
9. Hans-Otto Pörtner (Germany)
10. Debra Roberts (South Africa)
11. Joy Pereira (Malaysia)
12. Eric Brun (France)

13. Kevin Hennessy (Australia)
14. Gabrielle Pétron (Canada)
15. Eduardo Calvo Buenda (Spain)
16. Trigg Talley (United States of America)
17. Carlos Manuel Rodriguez (Costa Rica)
18. Myles Allen (United Kingdom of Great Britain and Northern Ireland)
19. Jiaqing Xue (China)
20. Naser Al-Khater (Qatar)
21. Sonja van Renssen (Netherlands)
22. Marianne Karlsen (Norway)
23. Dechen Tsering (Bhutan)
24. Diana Ürge-Vorsatz (Hungary)
25. Abdalah Mokssit (Morocco)
26. Rosina Bierbaum (United States of America)
27. Saleemul Huq (Bangladesh)
28. Ayman Shasly (Saudi Arabia)
29. Fatima-Zahra Taibi (Morocco)
30. Anne Olhoff (Denmark)
31. Young-Woo Park (Republic of Korea)
32. Hilda Heine (Marshall Islands)
33. Alfredo Fidel (Cuba)
34. Valerie Masson-Delmotte (France) - ex-officio member as a Co-Chair of Working Group I

It's challenging to get a precise count of the number of scientists in the IPCC Bureau since it comprises people with diverse backgrounds and knowledge. Although some Bureau members may possess scientific expertise, most are government officials or policy specialists. Nevertheless, it's crucial to recognise that the primary authors of the IPCC's scientific assessment reports are scientists and researchers from various countries.

It is pertinent to note that the IPCC is no longer investigating or looking at what the exact cause of global warming is and has concluded with high confidence that the warming of the climate system is unequivocal and that human activities, particularly the emission of greenhouse gases, are the main cause of this warming.

The IPCC, however, regularly reviews scientific knowledge on the causes and impacts of climate change as new research and scientific methods are developed. The IPCC conducts periodic assessments of the state of the science on climate change, with each assessment building on the previous one and incorporating new research findings and advances in scientific methods.

The IPCC uses a range of methods to regularly review the scientific knowledge on the causes and impacts of climate change as new research and scientific methods are developed. Some of these methods include:

1. Literature review: The IPCC conducts a comprehensive review of the scientific literature on climate change, including peer-reviewed research papers, reports, and other relevant publications.
2. Expert review: The IPCC engages with a broad range of experts from around the world, including scientists, policymakers, and stakeholders, to provide feedback on the latest scientific findings and identify areas where more research is needed.
3. Modelling: The IPCC uses computer models to simulate climate scenarios and assess the impacts of different greenhouse gas emission pathways on the climate system.
4. Scenario analysis: The IPCC develops a range of scenarios that explore possible future trajectories of greenhouse gas emissions, taking into account different socio-economic and technological developments.
5. Synthesis reports: The IPCC periodically publishes synthesis reports that draw together the latest scientific findings on climate change, providing a comprehensive overview of the state of knowledge on the subject.
6. Special reports: The IPCC also produces special reports on specific aspects of climate change, such as the impacts of global

warming of 1.5°C above pre-industrial levels or the role of land use in climate change.

7. Expert meetings and workshops: The IPCC convenes expert meetings and workshops to discuss and review the latest scientific findings on climate change, identify research gaps, and provide guidance on future research priorities.

After the scientific community reached a 97% consensus on climate change, it was not possible to obtain a list of all the peer-reviewed research papers that had undergone review.

The IPCC's assessments are widely recognised as the most thorough and authoritative summaries of climate science to date. However, it is important to understand that science is an ever-evolving discipline, and new evidence and viewpoints must be taken into account. Every new scientific finding either disproves an old theory or improves upon it in some way. Therefore, it is essential to view the science behind climate change with nuance and objectivity.

While the IPCC's assessments are generally accepted as the best analyses of current climate science, they should not be taken at face value. There may be various ways to interpret the data, and some researchers may disagree with the findings presented in the reports. The scientific method is an ongoing process of learning and discovery, and it is essential that scientists keep an open mind to new evidence and alternate points of view.

The possibility of healthy debate among scientists must also be recognised.

The IPCC's conclusions are acceptable at face value for those who do not object to the available evidence.

However, the findings should not be taken as gospel or treated as infallible. These differences are important and deserve an open and honest discussion.

It is important to acknowledge that the IPCC reports are not the only resources available to educate oneself on the topic of global warming. Many other reputable scientific organisations and institutions conduct research and publish reports on climate change.

Taking into account a wide range of perspectives and viewpoints is essential for developing a thorough understanding of the issue.

Ex-IPCC scientists who subsequently disagreed with the current scientific consensus have been the target of vicious attacks. However, personal attacks and other forms of intimidation have no place in a scientific discussion. There should be room for reasonable disagreements in the scientific community, and those disagreements should be acknowledged, discussed, and resolved in an open and honest manner.

Assessments by the IPCC are often taken at face value, making them difficult to dispute or reinterpret. The IPCC's evaluations come off as too dogmatic and prescriptive. The best kind of criticism is the kind that leads to improvements in both quality and accuracy. The IPCC's assessments can be criticised, and it is important to recognise that there is room for reasonable disagreements in the scientific community.

To effectively tackle the issue of climate change, it is important that policymakers keep an open mind and have access to accurate and up-to-date information on the state of the planet's climate. The IPCC plays an important role in providing this information, but it is essential to

4.2 National Aeronautics and Space Administration (NASA)

The National Aeronautics and Space Administration, or NASA, is best known for its space exploration missions, but the agency also conducts extensive research on Earth's climate system and its changes over time via a fleet of satellites that provide critical data on various aspects of Earth's climate. These satellites collect data on the planet's atmosphere, land, and oceans, including temperature, glasshouse gases, sea level, ice cover, and vegetation dynamics. This information assists scientists in analysing long-term patterns, tracking environmental changes, and developing climate models.

NASA develops and manages extensive databases and archives holding massive volumes of climate-related data. Scientists, politicians, and the general public can use these tools to examine climate trends, conduct research, and devise measures to mitigate and adapt to climate change. Open data encourages transparency, collaboration, and evidence-based decision-making.

NASA scientists use powerful computer models to mimic the Earth's climate system and forecast future weather patterns. Data from satellite observations, ground-based measurements, and historical climate records are all used in these models. NASA researchers may analyse the possible repercussions of climate change and refine their understanding of its origins and consequences by running simulations based on alternative glasshouse gas emissions scenarios.

NASA investigates the global carbon cycle, which is concerned with the flow of carbon between the atmosphere, oceans, land, and living creatures. Understanding carbon dioxide (CO_2) concentrations, as well as carbon sources and sinks, is critical for determining the drivers of climate change. NASA satellite observations give important information about CO_2 distribution and absorption by forests, oceans, and other natural processes.

NASA works collaboratively with other space agencies, research institutes, and organisations around the world to address climate change. The Joint Polar Satellite System (JPSS) and the Integrated Global Observing System (IGOS) are global efforts to coordinate Earth observations, share data, and enhance climate models.

NASA and the IPCC have distinct roles and approaches when it comes to studying and addressing climate change. Here are the key differences between the two:

1. **Mandate and Structure:**
    NASA: NASA is an independent U.S. government agency responsible for space exploration, aeronautics research, and Earth science. Its mission includes studying the Earth's climate system as part of its broader scientific objectives.

IPCC: The World Meteorological Organisation (WMO) and the United Nations Environment Programme (UNEP) established the IPCC as an intergovernmental body. Its primary purpose is to provide policymakers with objective, scientific assessments of climate change, impacts, and potential mitigation strategies.

2. **Focus and Scope:**
NASA: NASA conducts scientific research and observations of Earth's climate system using satellite missions, ground-based measurements, and climate modelling. Its focus is advancing our understanding of climate processes, studying long-term trends, and providing data and information for scientific research and public knowledge.

IPCC: The IPCC assesses scientific, technical, and socioeconomic information on climate change. It reviews and synthesises existing research, including published literature, to provide comprehensive reports on the state of knowledge. The IPCC's primary focus is on assessing the impacts of climate change, analysing mitigation options, and providing policymakers with relevant information to inform decision-making.

3. **Composition and Expertise:**
NASA: NASA comprises a diverse community of scientists, engineers, and researchers with expertise in various fields, including atmospheric science, remote sensing, climate modelling, and Earth system science. Its scientists contribute to climate research through data analysis, modelling, and satellite missions.

IPCC: The IPCC consists of thousands of scientists from around the world who volunteer their time to contribute to its assessments. These scientists are selected based on their expertise and represent a broad range of disciplines, including climate science, economics, social sciences, and other relevant fields.

4. **Assessment Process:**

   NASA: NASA conducts ongoing research and provides data and observations to the broader scientific community. Its findings contribute to the body of scientific knowledge on climate change and inform policy discussions. However, NASA does not produce formal assessments or policy recommendations.

   IPCC: The IPCC conducts periodic assessments, typically every 5-7 years, that involve a comprehensive review of published research on climate change. These assessments undergo extensive peer review and involve multiple rounds of expert and government review to ensure scientific rigour. The IPCC's reports include policy-relevant information and provide guidance to policymakers based on the available scientific evidence.

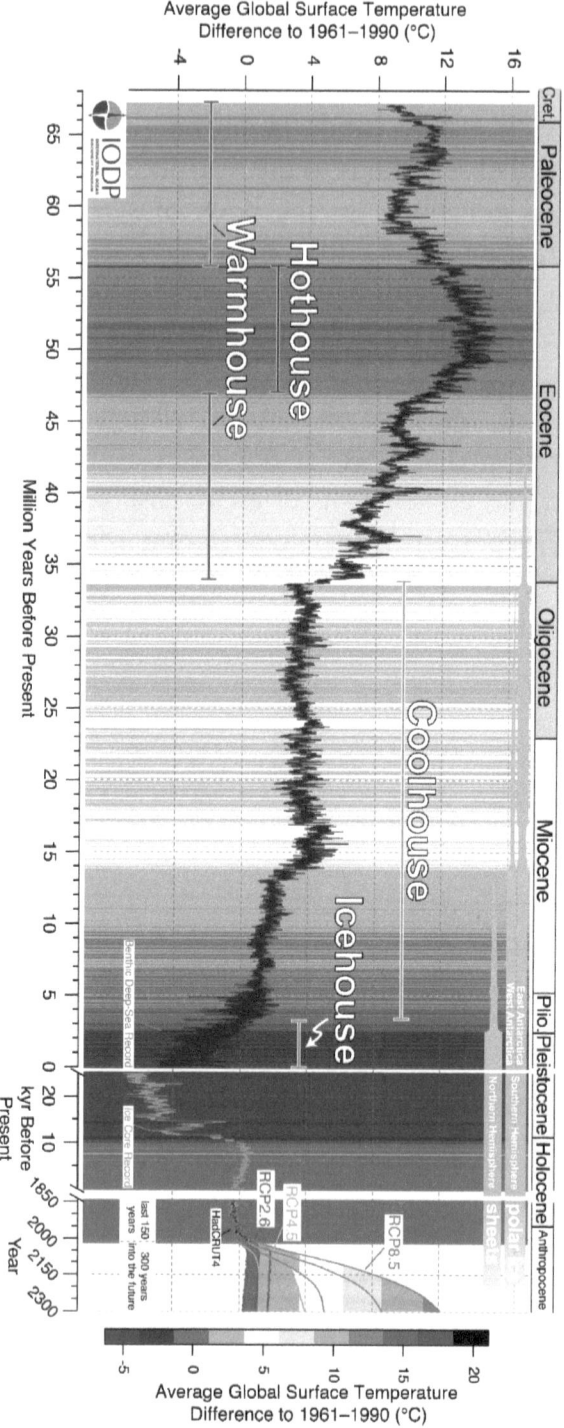

Figure 3: Average Global surface temperature between 1961-1990
Source: https://cdn.mos.cms.futurecdn.net/FdHtSNVNsWUbbbUqLMbEja.png

Collectively, both NASA and the IPCC have arrived at similar conclusions regarding climate change. Here are some key conclusions reached by these organisations.

Both NASA and the IPCC agree that climate change is occurring and that human activities, particularly the emission of greenhouse gases, are the primary drivers of the observed warming trends. They recognise that the increase in atmospheric concentrations of CO_2, methane, and other greenhouse gases is altering Earth's climate system.

NASA and the IPCC confirm that the Earth's average surface temperature has been increasing over the past century. They agree that the warming trend has been unprecedented in recent history and is largely due to human activities. The IPCC has stated that it is "extremely likely" (95-100% probability) that more than half of the observed increase in global average surface temperature (Figure 3) since the mid-20th century is attributed to human influence.

NASA and the IPCC acknowledge that climate change is causing significant impacts on natural systems. They observe changes such as shrinking ice caps and glaciers, rising sea levels, shifts in precipitation patterns, changes in ecosystems and biodiversity, and altered weather events. These changes have the potential to disrupt ecosystems, exacerbate water scarcity, threaten biodiversity, and affect agriculture and food security.

Both organisations highlight the risks and vulnerabilities associated with climate change. They recognise that certain regions and communities are more vulnerable to the impacts of climate change, particularly in developing countries and areas with limited adaptive capacity. They emphasise the need to address social, economic, and environmental vulnerabilities to build resilience and minimise future risks.

NASA and the IPCC emphasise the urgency of taking action to mitigate climate change and adapt to its impacts. They highlight the importance of reducing greenhouse gas emissions to limit global temperature rise and mitigate further climate change. Additionally, they emphasise the need to develop and implement adaptation strategies to cope with the changes that are already occurring and that are expected to continue in the future.

4.3 The Phoney Scientific Consensus

John Cook, Dana Nuccitelli, Sarah A. Green, Mark Richardson, Barbel Winkler, Rob Painting, Robert Way, Peter Jacobs, and Andrew Skuce (Cook et al., 2013) authored a seminal paper on the scientific consensus on climate change, which was published in the journal Environmental Research Letters. The paper analysed nearly 12,000 peer-reviewed scientific papers on climate change published between 1991 and 2011 and found that 97% of the papers that expressed a position on the issue endorsed the consensus view that human activities are causing global warming.

Another paper in 2016 by John Cook, co-authored by Naomi Oreskes, Peter Doran, William Anderegg, Bart Verheggen, Ed Maibach, J. Stuart Carlton, Stephen Lewandowsky, Andrew Skruce, Sarah Green, Dana Nuccitelli, Peter Jacobs, Mark Richardson, Barbel Winkler, Rob Painting, and Ken Rice (Cook et al. 2016), concluded that "the finding of 97% consensus in published climate research is robust and consistent with other surveys of climate scientists and peer-reviewed studies. The following paragraphs detail the credible credentials possessed by the researchers.

John Cook is an Australian climate communication researcher and the founder of the website Sceptical Science, which is dedicated to debunking climate change myths and misinformation. He is also an adjunct professor at the Global Change Institute at the University of Queensland, where he focuses on climate change communication. John Cook holds a PhD in cognitive psychology from the University of Western Australia.

Dana Nuccitelli is an environmental scientist and climate blogger based in the United States. He has written extensively on climate change for various media outlets, including The Guardian, and is known for his work on climate change denial. Dana Nuccitelli holds a bachelor's degree in physics from Reed College and a master's in environmental science from Western Washington University.

Sarah A. Green is a professor of chemistry and physics at Michigan Technological University and an expert in the study of atmospheric and environmental chemistry. Her research focuses on the chemistry of the atmosphere and the impact of human activities on air quality and climate. Sarah A. Green holds a bachelor's degree in chemistry from the University of Calgary and a PhD in physical chemistry from the University of British Columbia.

Mark Richardson is an assistant professor of atmospheric science at the University of Reading in the UK. His research focuses on understanding the physical mechanisms driving climate change, particularly the role of clouds and atmospheric radiation. Mark Richardson holds a bachelor's degree in physics from the University of Oxford and a PhD in atmospheric physics from the University of Edinburgh.

Bärbel Winkler is a research associate at the University of Cape Town in South Africa, where she focuses on the economics of climate change mitigation and adaptation. Her research interests include carbon pricing, emissions trading, and the design of effective climate policies. Bärbel Winkler holds a master's degree in economics from the University of Cape Town and a PhD in economics from the University of Maryland.

Rob Painting is a New Zealand-based climate researcher and writer who has written extensively on climate change for a variety of media outlets. His research focuses on the physical and chemical processes that drive climate change, particularly the role of the oceans in the Earth's climate system. Rob Painting holds a bachelor's degree in earth sciences from the University of Waikato and a PhD in atmospheric chemistry from the University of East Anglia.

Robert Way is a PhD candidate in the Department of Geography at the University of Ottawa in Canada, where his research focuses on the impacts of climate change on Arctic sea ice. He has also written extensively on climate change for a variety of media outlets. Robert Way holds a bachelor's degree in physics from the University of Guelph and a master's in geography from the University of Ottawa.

Peter Jacobs is a postdoctoral researcher at George Mason University in the US, where he studies climate change communication and the role of social media in shaping public perceptions of climate change. Peter Jacobs holds a bachelor's degree in geology from the University of Rochester and a PhD in climate science from the University of California, Santa Cruz.

Andrew Skuce is a retired Canadian computer scientist and climate blogger who has written extensively on climate change for various media outlets. He is known for his work on debunking climate change denial and communicating the science of climate change to the public. Andrew Skuce holds a bachelor's degree in mathematics from the University of British Columbia and a master's degree in computer science from the University of Waterloo.

> *"I'm a natural scientist. I'm out there every day, buried up to my neck in sh*t, collecting raw data, and that's why I'm so sceptical of these models, which have nothing to do with science or empiricism but are about torturing the data till it finally confesses."* Ian Plimer.

Ian Plimer is an Australian geologist and professor emeritus of earth sciences at the University of Melbourne. He has authored several books and articles on the subject of climate change, in which he argues that the scientific consensus on human-caused climate change is overstated and that natural factors play a larger role in climate variability than human activities.

Despite this consensus, forming the subject matter of the two research papers authored by a group of researchers briefed above, with the majority of the scientists conceding and agreeing to anthropogenic global warming, the rest (3%) are keeping the debate alive.

On the basis of the analysis of the signatories to the consensus document, it has been revealed that only 3% are climate experts, and the rest, 97%, do not meet the criteria of being climate scientists. Furthermore, most are involved with countermovement organisations or industries (Young, Laura D., and Erin B. Fitz, 2021). The consensus on human-caused climate change has its origins in public opinion and public perception.

It appears that Cook et al.'s overwhelming consensus (Figure 4) about anthropogenic global warming, published in a political rather than a scientific journal, is a public relations exercise disguised as a scientific method. It is a misuse of science and an abuse of statistics. It belittles the complexities of the subject of climate science. It is an effort to trivialise the issue, bring it down to the level of the belief system, and divide the scientists into believers and non-believers.

Ten years after the initial consensus, a survey of consensus among 2780 geoscientists found that 91 to 100 per cent agreed on anthropogenic global warming (Myers, Krista F., et al., 2021).

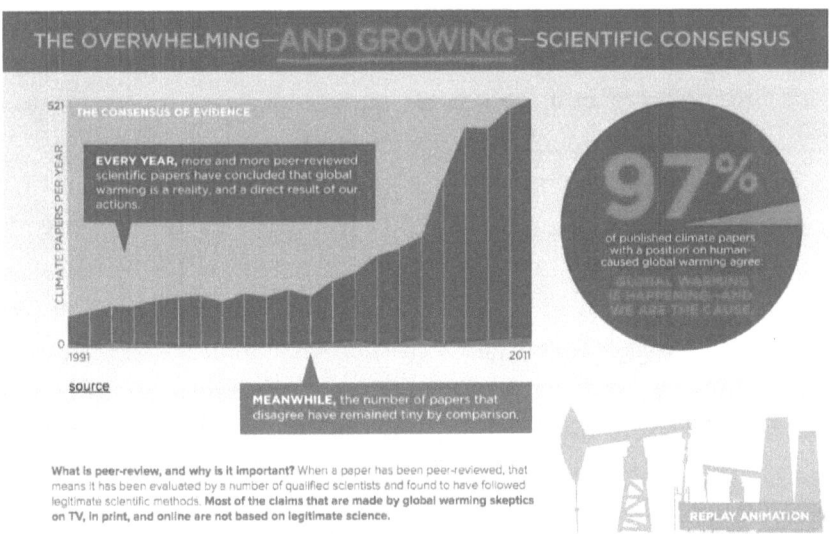

Figure 4: The Scientific Consensus
Source: https://i0.wp.com/www.skepticalraptor.com/blog/wp-content/uploads/2013/08/climate-change-consensus.jpg

"Let's be clear: the work of science has nothing whatsoever to do with consensus. Consensus is the business of politics. Science, on the other hand, requires only one investigator who happens to be right, which means that he or she has results that are verifiable by reference to the real world. In science, consensus is irrelevant. What is relevant are reproducible results. The greatest scientists in history are great precisely because they broke with the consensus." Michael Crichton, A.B. Anthropology, M.D. Harvard Michael Crichton was an American author, screenwriter, and film director best known for his science fiction and techno-thriller novels. Some of his most famous works include "Jurassic Park," "The Andromeda Strain," and "Timeline."

Human-induced climate change is a non-scientific, popular political view. The overall behaviour of the Earth system is strongly influenced by interactions among its various component systems, including the atmosphere, cryosphere, hydrosphere, oceans, pedosphere, lithosphere, and inner Earth, as well as by life and human activity.

The Earth System Dynamics (ESD) Journal provides a platform for sharing research and discoveries that enhance our comprehension of climate science and deepen our understanding of intricate environmental interconnections.

Earth System Dynamics (ESD) is a non-profit international scientific journal dedicated to the publication and public discussion of studies that take an interdisciplinary approach to the functioning of the Earth system as a whole and global change. ESD invites contributions that investigate various such interactions and the underlying mechanisms, methods for conceptualising, modelling, and quantifying these interactions, predictions of overall system behaviour in response to global changes, and the implications for its habitability, humanity, and future Earth system management by human decision-making.

A number of submissions to ESD claim that the increase in atmospheric temperature is due to anthropogenic activities.

The peer review process of those journals does not pass the ESD's test, on which their data is based, so the ESD has outright rejected these submissions. (Kleidon, A., et. al., 2023). These publications often include analyses from other research journals outside the climate community, presumably following a peer-review process. The reason for the rejection is that the editors of those journals lack scientific openness and knowledge of how the Earth's atmospheric greenhouse effect works.

In November 2009, there was a notable incident where a collection of emails and documents belonging to the Climate Research Unit (CRU) at the University of East Anglia was illicitly acquired and subsequently disclosed to the public. This event garnered substantial media coverage and stirred controversy, as certain individuals alleged that the leaked correspondence revealed unethical behaviour and manipulation of climate data by the implicated scientists.

The controversy had a significant impact on public opinion and the political discourse around climate change, with some commentators referring to it as "Climategate" and using it to cast doubt on the validity of climate science. Climategate had such an impact that it was said to have contributed to the failure of the United Nations COP15 climate change summit in Copenhagen, Denmark (Hunt, E., 2021). Climate change sceptics claimed that the emails revealed that scientists had manipulated data and suppressed dissenting opinions to promote the idea of human-caused climate change. Some critics of the CRU even accused the scientists of committing fraud.

4.4 The Geopolitics of Climate Change

The relationship between geopolitics and climate change is closely intertwined, as climate change is a global issue that has the potential to reshape global power dynamics. Geopolitics is the study of how geography, economics, and politics impact international relations and global power structures. The geopolitics of climate change encompasses the political, economic, and social implications of global warming, climate change, and efforts to address its effects.

Although climate change affects all nations and regions, its impacts are not evenly distributed, and the geopolitics of climate change are complex and multifaceted. This has significant implications for global politics, economics, and social systems. For example, countries that are major emitters of greenhouse gases may face diplomatic and economic pressure from other nations to reduce their emissions, which could lead to changes in global power dynamics as countries transition towards more sustainable and low-carbon economies.

Chaturvedi and Doyle emphasised that the global north, which benefits from polluting activities, also shapes the discourse around climate change as a means of geopolitical coercion against the global south, which is frequently unjustly accused of being responsible for climate change. (Chaturvedi, Sanjay, and Timothy Doyle, 2016).

Climate change is a global phenomenon, and emissions in one part of the world can affect the entire planet. The geopolitics of climate change is concerned with how efforts to mitigate and adapt to climate change will impact relationships between political jurisdictions and nation-states. This is a classic example of how climate change and international relations intersect and is a critical aspect of geopolitics.

In 1997, a mechanism was developed to require industrialised countries to commit to reducing their greenhouse gas emissions. This led to the adoption of the Kyoto Protocol on December 11, 1997, in Kyoto, Japan. However, it took approximately seven years for the protocol to come into force, which occurred on February 16, 2005.

The reason for the delay was due, in large part, to negotiations around the creation of a fund to help developing countries implement carbon reduction programmes, as proposed by Brazil.

Additionally, the US requested the ability to purchase carbon credits from developing countries with lower-cost mitigation programmes. Eventually, these two issues were resolved, leading to the successful implementation of the Kyoto Protocol.

The Copenhagen Summit, also known as the 2009 United Nations Climate Change Conference, took place in Copenhagen, Denmark, from December 7 to 18, 2009. On December 18, the United States, China, India, Brazil, and South Africa drafted the Copenhagen Accord, which recognised climate change as one of the most significant challenges facing the world today. The accord called for action to keep global temperature increases below 2 °C. However, it should be noted that this document was not legally binding and did not include any binding commitments to reduce carbon dioxide emissions.

According to Desai (Desai, N. 2012), despite the Copenhagen Agreement, it is unlikely that climate actions, impacts, and diplomacy will significantly change the current global power balance. According to Payne (Payne, R.A. 2007), vested interests in fossil fuel production and consumption may oppose proponents of green policies. This may hinder efforts to reach an agreement on collective action to mitigate the adverse effects of climate change and could even escalate into a geopolitical issue or conflict.

According to Branko Bosnjakovic (Bosnjakovic, Branko, 2010), there appears to be a shift in global power towards the East, which will have implications for the geopolitics of climate change. It appears that the United States has placed a higher value on maintaining its powerful position in the world, which is thought to rely on the ongoing usage of oil and gas, than on its policies regarding climate change.

Climate change governance is the set of institutions, policies, and methods used by the global community to address issues related to climate change. In contrast, geopolitics involves political and strategic considerations related to geography, power, and resources. The intersection of these two concepts results in the notion of the geopolitics of climate change governance.

The distribution of power and resources among different countries and regions is a crucial aspect of the geopolitics of climate change governance. In the past, the industrialised countries of the West have had the most influence on global climate change policy due to their large emissions. However, emerging economies such as China, India, and Brazil have also become significant players in the climate change arena in terms of emissions and shaping policy.

Another important factor in the geopolitics of climate change governance is the distribution of renewable energy resources. Countries with vast reserves of renewable energy, like solar, wind, or geothermal, have an advantage in the transition to a low-carbon economy. Consequently, there is competition for control over these resources and the technology and infrastructure required to harness them.

Furthermore, there is an increasing awareness of the geopolitical risks associated with climate change. These risks include the displacement of populations due to sea level rise or extreme weather events, the destabilisation of vulnerable states, and the potential for conflicts over scarce resources. Thus, global cooperation is necessary to tackle these challenges, but there is a risk that geopolitical tensions may escalate if countries cannot agree on how to share the costs and benefits of climate action.

Chapter 5

The Science

Science is a systematic and evidence-based approach to acquiring knowledge and understanding of the natural world. It involves the use of observation, experimentation, and analysis to test hypotheses and develop theories that explain the workings of the universe. It encompasses a broad range of disciplines, including physics, chemistry, biology, geology, astronomy, and many others. Scientists use various tools and methods to research and collect data, including laboratory experiments, field observations, computer simulations, and mathematical modelling.

It is an ongoing process, and discoveries and insights are constantly being made. Scientists publish their research in academic journals, and their findings are subject to peer review by other scientists in their field.

The scientific method, a vital element of science, involves a sequence of steps. Initially, scientists observe natural phenomena in order to discern patterns or relationships. They then formulate a hypothesis based on their observations to provide an explanation for the observed phenomena. Next, scientists use their hypotheses to generate predictions regarding future observations or experiments. To test their hypothesis, they design and carry out experiments that may either corroborate or disprove their predictions. Following the experimentation phase, scientists analyse the data they have collected to ascertain whether their hypothesis is supported or invalid. Finally, drawing on the outcomes of their experiments, scientists draw conclusions about the veracity of their hypothesis.

The scientific method, a vital element of science involving a sequence of steps, is instrumental in understanding the potential impacts of continued warming and exploring the different theories that explain the causes and consequences of climate change.

5.1 Theories of Climate Change

Theories of climate change provide an explanation for how and why the Earth's climate has changed over time, and they help us understand the potential impacts of continued warming. This chapter will explore the science behind climate change and the different theories that explain its causes and consequences.

As such, the subject of climate change is complex and multidisciplinary, and we need to take input from a vast number of scientific professions, including earth science, biological science, chemistry, physics, and a whole range of hybrid scientific groups like biochemistry, geochemistry, geochemistry, geophysics, and, of course, meteorology. But at the moment, the majority of the literature made available on the subject comes not from people with backgrounds in either of the fields listed herein but from behavioural sciences like sociology, psychology, and politics. The subject of climatology is still emerging. The advances in technology and communications have provided tools to scientists from all fields and enabled their reach at all levels.

There are many theories that do not claim global warming to be man-made (Bast, Joseph, 2010; Panofsky, Hans A., 1956; Hamilton, Clive, 2012). However, all of these theories are not mutually exclusive, and the phenomenon of climate change could be a consequence of different theories in different proportions as contributory sources.

Panofsky (Panofsky, Hans A., 2010) clubs climate warming theories into three categories: earth's crust, atmospheric, and astronomical. Bast, Joseph, op cit. (2010) lists seven theories of climate change: bio-thermostat, cloud formation and albedo, ocean currents, planetary motion, solar variability, human forcings, and anthropogenic. Hamilton, C. (2012), op. cit., in his review essay of the reviews of three books, discusses the three theories propounded by the authors in their books. However, all three books are solely concerned with anthropogenic climate change and approach the subject from different ideological perspectives.

Climate change is attributed to several natural processes that have contributed to alterations in global temperature, sea level, and weather patterns. These processes include solar irradiance, natural cycles, greenhouse gases, volcanic activity, plate tectonics, and cosmic rays. According to the solar irradiance theory, fluctuations in the sun's radiation output may influence Earth's climate by altering the amount of solar radiation that reaches the planet's surface. The natural cycles theory proposes that climate changes observed presently are part of a natural cycle and not primarily caused by human activity, as fluctuations have happened before due to natural factors such as changes in the Earth's orbit and axial tilt. The greenhouse gas theory suggests that increases in greenhouse gas concentrations such as carbon dioxide from human activities like deforestation and fossil fuel burning trap heat in the atmosphere, leading to the greenhouse effect and warming the Earth's climate. Volcanic activity can release gases and particles that cool the planet by reflecting sunlight. Plate tectonics can affect climate through changes in ocean currents and atmospheric circulation patterns caused by the movement of continents. Finally, the theory of cosmic rays proposes that changes in cosmic rays from outer space can affect cloud formation, ultimately affecting Earth's climate.

Climate change has become a growing concern for people around the world. The Earth's climate has been changing for millions of years. Scientists have been studying the causes of climate change for many years. There are various theories about the underlying mechanisms and factors that contribute to climate change. In this context, it is important to explore and understand the different theories of climate change to gain a comprehensive understanding of this global issue.

5.1.1 Bio-Thermostat Theory

In the field of climate science, a few ideas and models seek to explain how the Earth's climate system can maintain a stable temperature despite changes in numerous parameters, such as the quantities of greenhouse gases. One of these hypotheses is known as the "Bio-Thermostat Hypothesis," and it claims that there are negative feedback processes in the climate system of the Earth that work like a thermostat and keep the temperature of the entire planet stable within a relatively small range.

According to this idea, some aspects of the environment, such as cloud cover, ocean circulation, and ice cover, can all play a role in the formation of negative feedback that limits the degree to which temperatures can shift. The excess carbon dioxide that is released into the atmosphere (positive feedback) is balanced out (negative feedback) by biological and chemical processes such as carbon sequestration, carbonyl sulphide, diffusion of light, iodine compounds, and dimethyl sulphide, which together act as a bio-thermostat to maintain a constant temperature.

Plants take up most of their nutrition through the air, so increasing the amount of carbon dioxide in the atmosphere is beneficial. Through a series of biochemical processes known as "carbon sequestration," plants are able to store the extra carbon dioxide in the atmosphere in their leaves, branches, trunks, and roots, in addition to the soil beneath the plants.

Because of the biological processes in the plants and other vegetation, the soils naturally give off carbonyl sulphide (COS) gas. In addition, it can enter the atmosphere, where it undergoes a transformation into sulphate aerosol particles and is able to reflect solar energy back into the atmosphere. Because of the way in which this process works, the earth has a cooling effect, which helps to counteract the impact of any excessive levels of carbon dioxide in the atmosphere.

As higher concentrations of carbon dioxide promote more plant productivity, this enables plants to produce more gases that get converted into aerosols called biosols through a process called diffusion of light that, in turn, acts as nuclei for cloud production. In other words, higher concentrations of carbon dioxide promote greater plant productivity. Marine algae generate iodine, which leads to the formation of organic iodine compounds in the water, similar to how plants on land produce these chemicals. These molecules produce a cooling impact by acting as biosols and contributing to the production of clouds, which in turn reflect back the sun's energy. This counteracts the warming effect that carbon dioxide has on the atmosphere.

Oceans release a biological gas called dimethyl sulphide (DMS), which has a direct correlation to the temperature of the surrounding air. If the temperature is higher, the amount of DMS the seas produce will also be higher. DMS is a significant contributor to the production of clouds. When there are more clouds, they have a greater capacity to reflect the sun's rays.

5.1.2 Clouds and Albedo Theory

The "Clouds and Albedo Hypothesis" is a scientific theory that suggests that variations in cloud cover and the reflection of the Earth's surface (also known as albedo) can have a significant impact on the climate of the Earth. This theory was developed in the 1970s. The hypothesis states that as greenhouse gas emissions from humans cause the Earth to warm, this may produce changes in cloud cover and albedo, both of which can either magnify or moderate the warming effect.

Clouds play a multifaceted and ever-changing role in the climate system of the Earth. They are responsible for reflecting sunlight back into space and retaining heat in the atmosphere. Clouds that are low to the ground, like stratus clouds, have a tendency to have a cooling influence on the climate of the Earth because they reflect more sunlight back into space than the heat they contain. Since they absorb more heat than they give off, high clouds like cirrus tend to have a warming effect on the surrounding environment. Alterations in cloud cover and position can have a sizable effect on the amount of sunlight reflected back into space, which in turn can have an effect on the energy balance of the planet and, ultimately, its temperature.

Alterations in the reflectivity of the surface of the Earth, commonly referred to as "albedo", can also have an impact on the energy balance of the planet. For instance, the melting of ice and snow can lower the albedo of the earth. This is because the darker surface exposed as a result of this process absorbs more sunlight, which warms the surrounding area. This can result in a feedback cycle in which the melting of ice leads to additional warmth, which leads to more melting, and so on.

Although there is evidence to support the "Clouds and Albedo Hypothesis", it is important to note that the impact of clouds on the climate of the Earth is a complex and ongoing area of research. Furthermore, the precise nature and magnitude of their effect on the climate of the Earth are still the subject of much scientific debate and investigation.

The idea suggests that a rise in global temperatures could lead to a reduction in the amount of snow cover and sea ice, lowering the Earth's albedo and causing the surface to absorb more sunlight. This could, in turn, lead to further warming as well as a reduction in cloud cover, both of which could further worsen the trend of warming.

According to this idea, alterations in the production and albedo of clouds provide negative feedback that cancels out and balances out the warming influence that carbon dioxide has on the climate. Sud, Y. C., et al. (1999) found that variations in cloud cover in tropical regions served as a thermostat to maintain a surface temperature range between 28 and 300 degrees Celsius in the oceans. The examination of these data demonstrates that as the temperature of the sea surface rises, the air at the base of the clouds becomes charged with the moist static energy required for clouds to rise into the upper troposphere. Once there, they reduce the amount of solar radiation that reaches the sea surface, thereby producing a cooling effect. This phenomenon, which acts like a thermostat, contributes to the ventilation of the ocean and helps to keep the temperature of the sea surface within the range of 28–300 °C (Sud, Y. C., et al., op. cit.).

In a comparable manner, an examination of data on upper-level cloudiness and the mean sea-surface temperature in cloudy regions uncovered a significant inverse correlation between upper-level clouds and the mean sea-surface temperature in the eastern Pacific (Lindzen et al., 2001). According to their theory proposed in 2001, this phenomenon suggests that the dense, cloudy region acts as an adaptive infrared iris, ensuring stability in the temperature of the sea surface. Subsequent satellite data provided further evidence supporting the existence of this adaptive infrared iris phenomenon (Spencer et al., 2007).

5.1.3 Ocean Current Theory

According to the "Ocean Currents Hypothesis", fluctuations in ocean currents may have the potential to have a considerable influence on the climate of the Earth. Due to their dependence on variations in temperature, salinity, and density, ocean currents are crucial for regulating the Earth's climate and moving heat across its surface.

Changes in ocean currents, such as the Gulf Stream, which brings warm water from the Gulf of Mexico to the North Atlantic, are thought to have the potential to influence the climate of the Earth in a number of different ways. For instance, if the Gulf

Stream were to slow down or stop completely, as some climate models have indicated, this may contribute to a cooling effect in the North Atlantic and certain portions of Europe. This is because, in comparison to other regions at the same latitude, the warm water that the Gulf Stream typically transports helps keep these areas relatively mild.

Other ocean currents, such as the Antarctic Circumpolar Current, which helps to distribute heat around the Southern Ocean, and the Pacific Equatorial Countercurrent, which helps to regulate El Nino and La Nina events, also play important roles in the process of regulating the climate of the Earth, in addition to the Gulf Stream, which is the most prominent of these other ocean currents.

Although there is proof for the Ocean Currents Theory, it is crucial to remember that changes in ocean currents are complex and subject to various influences. These factors include shifts in atmospheric temperature, changes in wind patterns, and freshwater input from melting ice and precipitation. Although there is proof for the Ocean Currents Theory, it is important to remember that a variety of factors can affect changes in ocean currents. As a result, there is still a great deal of dispute and enquiry going on in the scientific community over the precise nature and magnitude of their effect on the climate of the planet.

Changes in ocean currents, according to this idea, might potentially lead to changes in the distribution of heat around the globe, which, in turn, could impact weather patterns and global temperatures. Changes in the El Nino Southern Oscillation (ENSO) in the Pacific Ocean could potentially have far-reaching effects on weather patterns around the world. For instance, a slowing or redirection of the Gulf Stream in the Atlantic Ocean could potentially lead to a cooling of the North Atlantic region. On the other hand, changes in the Gulf Stream in the Atlantic Ocean could potentially lead to a warming of the South Atlantic region.

Ventilation refers to the process through which water in the ocean continuously percolates to greater depths. This process is ongoing.

It takes one to two thousand years for the entire process to fully ventilate itself. The salty, cold water that originates in the polar regions eventually makes its way to the warmer, less salty water in the tropics. This deep ocean circulation, known as the Meridional Overturning Circulation (MOC), happens in two parts: the primary Atlantic Thermohaline Circulation (THC) and the secondary Surrounding Antarctica Subsidence (SAS) (Gray, William Mulroy, 2009). Both of these parts take place in the Atlantic Ocean. Because of the way the ocean's surface winds are configured horizontally, in addition to the global ocean's overall cover, there are also some up-and-down ocean currents that make up the Ekman pattern. Depending on where they occur, these changes may also contribute to local or global temperature variations.

5.1.4 Planetary Motion Theory

In the field of climate research, there are a few models that make an effort to comprehend how shifts in the Earth's orbit and spin can have an effect on climatic patterns over extended stretches of time.

The Milankovitch Theory is one example of this type of concept. This theory proposes that variations in the Earth's orbit and rotation can influence the distribution of solar radiation over the globe, which in turn can affect the climate of the planet. According to this idea, shifts in the Earth's orbit and tilt can cause shifts in the amount and distribution of solar radiation that reaches the Earth's surface, which in turn can drive long-term changes in climate patterns. These shifts in climatic patterns can have an effect on human societies.

According to this theory, the solar system's natural gravitational and magnetic oscillations—a result of the planet Earth's motion through space—cause climatic fluctuations. The path that Earth takes around the sun is elliptical. The point in its orbit when it is closest to the sun is called perihelion, and the point when it is farthest away from the Sun is called aphelion. Every 22,000 years, the date of the perihelion shifts due to a phenomenon known as the precession of the equinoxes. Because of the gravitational pull of the other planets, the eccentricity of the Earth shifts on timescales ranging from 100,000 to 400,000 years. On a cycle that lasts 41,000 years, the axis around which the Earth rotates alternates between lower and higher places. The shifts in the patterns of these motions are the cause of the changing climate.

5.1.5 Solar Variability Theory

The "Solar Variability Theory" theory originated in the 1970s, proposing that changes in solar activity might impact the quantity of energy reaching the Earth's surface, consequently influencing the planet's temperature and overall climate. This theory suggests that alterations in solar activity, including fluctuations in the sun's radiation strength and the number of sunspots, can exert notable influences on Earth's climate. Changes in solar activity have been shown to influence Earth's temperature in the past. During the "Maunder Minimum" in the 17th century, when there were few sunspots, Europe endured a period of abnormally cold winters known as the "Little Ice Age".

This theory contends that solar phenomena like sunspots and solar winds cause temperature changes. Sunspots are bursts of intense particles and radiation caused by changes in the sun's brightness. These recurring periods occur at 11, 87, and 210-year intervals. Electromagnetic radiation (or "solar wind") is subject to variations as a result of these cycles. These changes affect cloud formation and thermohaline circulation (TMC), both of which have effects on weather and climate.

Electrons from space-borne cosmic rays have a role in the formation of atomic-scale clusters of sulfuric acid and water nuclei. Reduced cloud cover means a less atmospheric reflection of the sun's heat and, thus, higher temperatures (Svensmark, H., 1997). The force of solar winds increases as magnetic activity on the sun rises. Together, these two variables block the entry of cosmic rays into the atmosphere, resulting in fewer clouds.

As a hypothesis in climate science, the Solar Variability Theory suggests that shifts in the sun's total energy output may significantly affect weather patterns here on Earth. The sun is the primary source of energy that drives Earth's climate system; hence, changes in the sun's supply of energy can affect global temperature trends.

This theory suggests that changes in solar activity, such as the frequency with which sunspots appear or the strength of solar flares, might affect the amount of energy that reaches Earth's atmosphere. Climate change over the past several millennia has been seen, and scientists have speculated that differences in solar activity may be to blame.

Despite the fact that this is true, it is crucial to keep in mind that scientists are still debating the hypothesis of solar variability. However, many studies have found that the effects of solar variability are relatively small when compared to the effects of other factors like the concentrations of greenhouse gases and aerosol emissions on the Earth's climate.

5.1.6 Human Forcings Besides Greenhouse Gases

According to this theory, human activities such as deforestation, urbanisation, and desert irrigation are more responsible for climate change than greenhouse gas emissions. Climate scientist Roger Pielke, Sr., from the University of Colorado at Boulder, believes that human influences are significant and involve a wide range of climate forcings, including carbon dioxide. Other "human forcings" include the urban heat island effect, nanoparticles and ozone, deforestation, growth in coastal areas, and rocket trails. The climate impacts of some of these "human forcings" can be comparable to or even greater than those of anthropogenic greenhouse gas emissions in specific regions. Therefore, the theory of anthropogenic global warming (AGW) can account for very little warming. However, the IPCC in 2007 did not acknowledge the importance of these other human climate forcings in altering regional and global climate, and it relied too much on a small subset of human-caused climate forcings to characterise the global climate as a whole (Pielke Jr., R. 2005).

5.1.7 Anthropogenic Theory

The idea of anthropogenic climate change was mooted and hypothesised in the 1890s by Svante Arrhenius (19 February 1859–2 October 1927), a Nobel laureate in 1903 in chemistry (Abatzoglou, J., et al., 2007). When people talk about the theory of climate change in everyday conversation, they often refer to the anthropogenic theory. This is the widely known and discussed theory that suggests human activities are causing the planet to warm up through the greenhouse effect, where certain gases like carbon dioxide trap heat in the Earth's atmosphere. It's important to note that while human-generated carbon dioxide emissions from activities like burning fossil fuels and industrial processes contribute to greenhouse gas levels, they make up only a small portion of the total amount released since the Industrial Revolution began with the invention of steam engines in 1760.

Industrial activities during the Industrial Revolution and other human activities integral to modern society, such as burning fossil fuels, mainly in power plants and automobiles, have released carbon dioxide, which has trapped more heat in our atmosphere for over 200 years. Besides carbon dioxide, methane and nitrous oxide are released into the atmosphere due to human activity.

A major part of the greenhouse gases is water vapour (36–90%), followed by methane (4–9%), ozone (3–7%), and carbon dioxide (1-6%). These percentages are rough, contested, and controversial figures.

Currently, the carbon dioxide level is 390 ppm and has increased by one-third in the last 200 years (MacFarling Meure, C., et. al.,

2006). The rate of increase in carbon dioxide during the abrupt global warming 183 million years ago and 55 million years ago was broadly similar to what we have today (Cohen, A.S., Coe, A.L., and Kemp, D.B., 2007).

Over 1.85 trillion metric tonnes of carbon dioxide (500 billion metric tonnes of carbon) have been added to the atmosphere due to human activities, 65% of which came from the burning of fossil fuels (Houghton, J., 2009; Andrés, R.J., Marland, G., Boden, T., and Bischoff, S., 2000; Anon, Anon, 2010; Metz, B., Davidson, O., et al., 2007).

It is estimated that humans have released approximately 2.4 trillion metric tonnes of carbon dioxide (CO2) into the atmosphere since the start of the Industrial Revolution in the mid-18th century. This figure is based on data from various sources, including historical records of fossil fuel consumption, industrial processes, and changes in land use, among others.

You may see the above two paragraphs as contradictory. The first paragraph states that over 1.85 trillion metric tonnes of carbon dioxide have been added to the atmosphere due to human activities, with 65% coming from burning fossil fuels. The second paragraph estimates that humans have released approximately 2.4 trillion metric tonnes of carbon dioxide into the atmosphere since the start of the Industrial Revolution.

It is important to distinguish between the total amount of carbon dioxide released by humans since the start of the Industrial Revolution and the cumulative amount of carbon dioxide released into the atmosphere as a result of human activities up until a specific point in time, as was mentioned in the second paragraph.

The figures in the first paragraph may refer to a more recent timeframe, while the second paragraph covers a longer period of time. Additionally, the first paragraph may be specifically referring to carbon dioxide emissions from fossil fuels, while the second paragraph includes emissions from a wider range of sources, such as land use changes and industrial processes.

Friedlingstein, P., et al. (2020) provides an updated estimate of cumulative carbon dioxide emissions from human activity between 1750 and 2018 based on data from fossil fuel consumption, cement production, and land use changes. Le Quéré (Quéré, C. L., et al. 2014) provides an estimate of cumulative carbon dioxide emissions from human activity between 1750 and 2014 based on data from fossil fuel consumption, cement production, and land use changes. Joos (Joos, F., et al. 2013) provide an estimate of cumulative carbon dioxide emissions from human activity between 1750 and 2011 based on data from fossil fuel consumption, cement production, and land use changes. Boden et al. (2017) provide an estimate of cumulative carbon dioxide emissions from fossil fuel consumption between 1751 and 2014 based on historical energy consumption data.

Plants and trees play an essential role in the carbon cycle regulating the climate because they absorb carbon dioxide from the air and release oxygen into the atmosphere. Forests and bushland act as carbon sinks and are a valuable means of maintaining the temperature. But humans clear vast areas of vegetation around the world for farming, urban and infrastructure development, or timber. When vegetation is removed or burned, the stored carbon is released into the atmosphere as carbon dioxide, contributing to global warming. Humans, animals, and livestock like sheep and cattle produce methane, a greenhouse gas. When livestock graze on a large scale, as in Australia, the amount of methane produced is a big contributor to global warming.

Another potent greenhouse gas that farming practices produce is nitrous oxide. Nitrous oxide is released during commercial and organic fertiliser production and use. Nitrous oxide also comes from burning fossil fuels and burning vegetation and has increased by 18% in the last 100 years.

Chlorofluorocarbons (CFCs) are chemical compounds of entirely industrial origin. They were used as refrigerants, solvents (substances that dissolve others), and spray-can propellants.

An international agreement known as the Montreal Protocol now regulates CFCs because they damage the ozone layer. Despite this, emissions of some types of CFC spiked for about five years due to violations of the international agreement. Once members of the agreement called for immediate action and better enforcement, emissions dropped sharply, starting in 2018.

Cook et al. used the Web of Science science-citation research site to review the titles and abstracts of peer-reviewed articles from 1991–2011 with the keywords "global climate change" and "global warming." They classified the articles into seven categories, from 'explicit endorsement with quantification' to 'explicit rejection with quantification," with "no position" in the middle. This erroneous and subjective approach is applied with the assumption that the publishing scientists who accept a theory will say so in the title or abstract. To count an article as part of the consensus, Cook et al. required that it "address or mention the cause of global warming." Of the 11,944 articles that came up in their search, 7,970—two-thirds—did not commit to either side. Cook et al. classified those articles as taking no position and thus ruled them out of the consensus.

It is worth noting that John Cook is not a climate scientist but a cognitive scientist. Cognitive science is an interdisciplinary field that studies the nature of the human mind, including how people perceive, think, reason, and remember. It encompasses a wide range of fields, including psychology, neuroscience, linguistics, philosophy, computer science, and anthropology. Cognitive scientists seek to understand how humans acquire, process, and use information and how cognitive processes are related to behaviour. They use a variety of methods and techniques to study these processes, including brain imaging, behavioural experiments, and computational modelling. That is why he has been able to spin the falsehood of a scientific consensus on climate warming.

Among the 928 abstracts of peer-reviewed research papers on anthropogenic global warming published between 1993 and 2003, none disagreed with the consensus position on climate change (Naomi Oreskes 2004). According to a survey of earth scientists, human activity in global warming is nonexistent. Michael Crichton (2004), in his book 'State of Fear, emphasises that the public's catastrophic climate is being manipulated through the media by creating fear psychosis.

The examination of anthropogenic theory delves into the extensive scientific consensus while acknowledging the presence of dissenting views from scientists who challenge the prevailing understanding of human-induced climate change. The conclusions drawn by these individuals question the widely accepted call for scientific consensus on the subject. Here are a few examples:

Richard Lindzen, a retired atmospheric physicist and professor at the Massachusetts Institute of Technology (MIT), has expressed scepticism, cited several times elsewhere in this document, about the degree to which humans are responsible for climate change and has suggested that the climate is less sensitive to greenhouse gases than many other scientists believe.

Judith Curry, a former climatologist and professor at the Georgia Institute of Technology, has raised questions about the degree to which human activities contribute to climate change and has suggested that the scientific agreement on the topic may be exaggerated. Over the course of her career, she has authored multiple research papers on climate change. In Curry, J. A., and Webster, P. J. (2011) 2011 book, "Climate Science and the Uncertainty Monster", Curry and Webster contend that uncertainty is an essential aspect of climate science that should be conveyed more clearly to both policymakers and the general public.

Judy Curry's 2008 paper, "Dynamic Analysis of Weather and Climate Regimes: Transitioning Between Attractors", delves into the dynamics of climate regimes and introduces a novel methodology for analysing the shifts between them. Her 2011 publication, "The Uncertainty Monster at the Climate Science-Policy Interface," tackles the difficulties of effectively conveying uncertainty in climate science to policymakers and the general public. In "Climate Change and Trace Gases" (2013), Curry examines the impact of trace gases, such as methane and carbon dioxide, on climate change and evaluates the evidence for human involvement in their atmospheric concentrations. Finally, in "Climate Models for the Layman" (2014), she presents a basic overview of the structure and function of climate models as well as their utility in comprehending the complexities of the climate system.

It's worth noting that several of Judy Curry's papers have had their links removed, likely due to individuals or organisations with their own agendas. The papers in question include "Combining hurricane hazards with climate change: implications for Policy" (Curry, Webster, Holland, and Pielke Jr., 2006), which is available on ResearchGate; "Climate Science: irreducibly uncertain" (Curry, 2017), available on Nature.com; "Climate models for the layman" (Curry, 2018), available on AMS journals; and "Climate change: the 2018 report of the Lancet Countdown on health and climate change" (Curry and Emanuel, 2019), which is available on The Lancet.

William Happer, a retired physicist and former professor at Princeton University, has expressed doubt about the extent to which carbon dioxide emissions contribute to climate change and has suggested that higher carbon dioxide levels could have beneficial effects on the environment. Happer has published research papers on various topics related to atmospheric physics and climate change. In his 2015 paper, "The Myth of Carbon Pollution", he argues that "carbon pollution" is a political term with no scientific basis and that increased levels of carbon dioxide may have positive effects on the environment. His 2017 paper, "Physics, Radiative Transfer, and Climate," provides an overview of the physical principles underlying climate models and argues that these models are unable to predict future climate change accurately. In his 2018 paper, "Carbon Dioxide and the Earth's

Future: Pursuing the Prudent Path", Happer asserts that carbon dioxide emissions are not a significant driver of climate change and that policies aimed at reducing emissions may be misguided and harmful.

In contrast, Willie Soon, an astrophysicist, has published research papers exploring the effects of solar radiation on climate change. In his 2001 paper, "Reconstructing Temperature Variations in the Bottom Water of the South China Sea Over the Last Glacial Cycle", Soon analysed sediment cores from the South China Sea to reconstruct past temperature variations. His paper argues that the region underwent significant temperature changes during the last glacial cycle, with the coldest period occurring about 20,000 years ago. In his 2003 paper, "Solar forcing of global climate change since the mid-17th century", Soon argued that solar radiation was the primary driver of global climate change since the mid-17th century. However, this paper received criticism for using flawed statistical methods and failing to account for other factors, such as greenhouse gas emissions. In his 2007 paper, "Variations in solar luminosity and their effect on the Earth's climate", Soon argued that variations in solar activity were the main driver of global climate change. His 2011 paper, "Re-evaluating the role of solar variability on Northern Hemisphere temperature trends since the 19th century," supports the idea that solar variability has played a significant role in Northern Hemisphere temperature trends since the 19th century.

Climatologist John Christy from the University of Alabama in Huntsville has contested the notion that climate models accurately predict global warming, instead suggesting that natural factors may contribute significantly to observed climate change. In "Satellite temperature record confirms warming trend" (2007), published in Nature, Christy analysed satellite data to show that the Earth's lower atmosphere had warmed by approximately 0.2 degrees Celsius per decade since 1979, a finding that supported previous research based on weather balloons and surface measurements.

In "What Do Observed Diurnal Temperature Ranges Tell Us?" (2008), published in the Journal of Geophysical Research, Christy studied diurnal temperature ranges (DTR) to explore the impact of global warming on temperature variability, finding that DTR had decreased globally, indicating that global warming was reducing temperature variability.

In his paper "Reconciling Observations of Global Temperature Change" (2010), published in Nature, Christy compared different temperature datasets to investigate disagreements among them, concluding that natural climate variability could explain some of the differences and that some datasets had overestimated warming trends while others had underestimated them. Additionally, in "Climate Models and Their Evaluation" (2018), published in Advances in Atmospheric Sciences, Christy reviewed the performance of climate models in predicting global temperature changes over several decades. She found that most models overestimated warming trends and did not reproduce observed temperature variability, highlighting the need for better model evaluation and more reliance on observational data.

Patrick Michaels, a former climatologist, has been critical of the scientific consensus on climate change and has argued that the impacts of climate change may be overstated.

According to Nir Shaviv, an astrophysicist at the Hebrew University of Jerusalem, variations in solar activity may be a more significant driver of climate change than the scientific community typically acknowledges, and carbon dioxide's contribution to global warming may be less than previously thought. In his 2002 paper, "The cosmic ray climate link: Evidence from the Past", published in the journal Advances in Space Research, Shaviv provided evidence for a link between cosmic rays and climate over long timescales. His 2005 paper, "On climate response to changes in the cosmic ray flux and radiative budget", published in the Journal of Geophysical Research, used climate models to investigate the impact of cosmic rays on climate and found that they could cause minor changes in global temperature but were unlikely to be the main driver of recent warming trends. In his 2008 paper, "Using the oceans as a calorimeter to quantify the solar radiative forcing", published in the Journal of Geophysical Research, Shaviv estimated the radiative forcing of the sun by measuring ocean heat content and found that solar radiation had contributed to global warming in the past but could not account for recent warming trends observed since the mid-twentieth century. His 2015 paper, "The human fingerprint in atmospheric composition changes during the Anthropocene", published in the journal Reviews of Geophysics, reviewed the evidence for human influence on the Earth's atmosphere and concluded that human activities such as fossil fuel burning and land use change had significantly impacted the composition of the atmosphere, resulting in increased greenhouse gas concentrations and other changes.

Roy Spencer, a climatologist at the University of Alabama in Huntsville, has contended that natural factors could be accountable for a considerable portion of observed climate change and that the effects of climate change might be overstated.

In his 1990 paper, "Global temperature monitoring with satellite thermal measurements", published in the journal Science, Roy Spencer introduced a new technique for monitoring global temperatures employing satellite thermal measurements. The study revealed satellite data could provide more precise temperature measurements than ground-based instruments.

In his 2007 paper, "An Inconvenient Truth: The Absence of Evidence for the Link between Carbon Dioxide and Global Warming", published in the journal Energy & Environment, Roy Spencer claimed that the proof for a causal relationship between carbon dioxide emissions and global warming was inadequate. The research scrutinised the reliability of climate models and suggested that natural climate variability could account for observed warming trends.

In his 2011 paper, "On the Misdiagnosis of Surface Temperature Feedbacks from Variations in Earth's Radiant Energy Balance", published in the journal Remote Sensing, Roy Spencer used satellite data to study the role of feedback mechanisms in climate change. The researchers discovered that the feedback response to warming was weaker than projected by climate models, indicating that the climate system was more resilient than previously assumed.

In his 2018 paper, "The Impact of Recent Forcing and Ocean Heat Uptake Data on Estimates of Climate Sensitivity", published in the journal Journal of Climate, Roy Spencer employed climate models to assess the sensitivity of the climate system to greenhouse gas emissions. The study found that climate sensitivity was lower than previously thought, suggesting that future warming trends might not be as severe as some climate models have predicted.

President Obama's May 16, 2013 tweet claimed that 97% of scientists agreed that climate change is real, man-made, and dangerous. However, scientific principles cannot be determined based on public perception or consensus, as in political or social issues. The source of the 97% consensus claim comes from an article titled "Quantifying the Consensus on Anthropogenic Global Warming in Scientific Literature" by John Cook et al. (2013) in Environmental Research Letters 2013, which reported that "Among abstracts expressing a position on AGW, 97.1% endorsed the consensus positions that humans are causing global warming". Despite this, recent global warming is not extraordinary when compared to data from the geological record. A meteorite strike 66 million years ago in Mexico rapidly cooled the Earth for a decade before warming it for nearly 10,000 years (Lear, Caroline H., et al., 2020) due to carbon dioxide released from marine carbonate evaporation.

Greta Thunberg, at the age of 15, is cited as an example of how those who promote the theory of climate change can be insensitive to others by using children to advance their personal agendas. This can potentially endanger the future of young people and contribute to the emergence of a new phenomenon known as "eco-anxiety", which is the feeling of worry, fear, and stress caused by the existential threat of climate change. Children and young people are particularly susceptible to experiencing climate anxiety, as they are often acutely aware of the potential long-term impacts of climate change on their future.

The issue of climate anxiety in children is becoming increasingly worrisome, as many kids feel helpless and hopeless in the face of the global climate crisis. The severity of anxiety and stress related to climate change can vary depending on individual circumstances and personal experiences. Climate alarmism, or the tendency to exaggerate or overstate the potential impacts of climate change, can contribute to the development of climate anxiety in children. It is crucial to recognise that these issues can be interconnected and may impact each other, leading to compounding effects on anxiety and stress.

Concerns have been raised about the potential psychological impact of activism on children, including the risk of burnout or emotional distress. It is essential to ensure that children are not exploited or put in dangerous situations. Some people have accused Greta Thunberg's parents and the adults around her of exploiting her for their own agenda. Additionally, there is growing research on how the tangible effects of climate change, combined with increased global calls to action, affect people's sense of humanity and self, resulting in emotional exhaustion in those concerned about climate change.

There is a debate over the concept of "climate alarmism", which some argue is false and unjust. The world is better today than in 1900, and according to the UN Climate Panel's middle-of-the-road projection for the end of the century, we will be even better off. However, governments are making irrational commitments as a result of climate alarmism, such as achieving carbon neutrality by 2050, which could cause a single-minded focus on climate change to overshadow crucial investments in health, education, jobs, and nutrition.

We must proceed with caution in order to prevent a single-minded focus on climate change from crowding out critical investments in health, education, jobs, and nutrition.

According to an American Psychological Association report on "climate grief" and a growing body of research on how the tangible effects of climate change, combined with increased global calls to action, affect people's sense of humanity and self, resulting in emotional exhaustion in those concerned about climate change, according to researchers at the University of Bath and members of the Climate Psychology Alliance (CPA) in the United Kingdom (Fearnow, B., 2019), the number of children being treated for "eco-anxiety" caused by climate change alarmists has increased because it is fueling fears that humans will become extinct as a result of their actions.

Climate alarmism is both false and unjust, as the world is far better today than in 1900 (Lomborg, B., 2020). According to the UN Climate Panel's middle-of-the-road projection for the end of the century, we will be even better off, with virtually no one remaining in extreme poverty and everyone much better educated. Rising sea levels are also a fabricated charge. According to the United Nations and the world's only Nobel laureate in climate economics, global warming will reduce the 21st century's welfare increase from 450% to "only" 434% of today's income. However, this is not an existential threat, and the media's portrayal of climate change's effects is frequently misleading.

Boute et al. (Boute, Anatole, and Alexey Zhikharev, 2019) label it vested interest insisting, ensuring, and advocating by the Russian government on developing and obtaining the competence and efficiency of local industry to undertake renewable energy projects, obtaining the available subsidies for the purpose, and avoiding big overseas transnational corporations. It is strange and criminal to canvass, stand, and advocate for local welfare and development rather than worrying about other corporate interests. It can be easily assumed that the authors are acting on behalf of and lobbying for some corporations as part of an underlying, hidden agenda.

The authors of this paper are raising an objection as to why a local industry group, Renova, which is also involved in the traditional energy field, is being allowed to develop and implement renewable energy projects and become a clean energy player at the cost of blocking the entry of foreign investors by legislating and provisioning the stringent requirements of local content in renewable energy projects. Apparently, the authors have declared the receipt of a research grant from the Research Grants Council of Hong Kong for this paper.

5.2 Geological Perspective on Climate Change

There are several questions to consider and reflect upon regarding climate change before arriving at any conclusions. Can anyone, including politicians, alter the course of climate change for better or worse? Can you comprehend the time frame of climate change, ranging from hours and days to years, centuries, millennia, and even millions of years in geological time? Are you knowledgeable about the origins of Earth, the universe, and life on Earth, as well as Charles Darwin's theory of evolution? Do you understand why certain species, such as dinosaurs, became extinct and why many other plants and animals are facing extinction today?

Are you aware that Homo sapiens is still undergoing evolution, and what changes may occur in the future? Have you considered the significance of discovering marine fossils at high elevations in the Himalayas? Are you aware of the continuing movement of the Earth's continents, known as "continental drift" (illustrated in Figure 6), and when this process began long before the emergence of the human species? It is necessary to find answers to these questions in order to arrive at a conclusion about the factors behind climate change and attribute responsibility for global warming to humans.

Are you aware that the Earth has experienced cycles of warming and cooling in the form of intermittent ice ages? Knowing this, it may be difficult to believe that humans are responsible for climate change or that humans have the ability to alter the direction in which it is headed. Where does all this fit in while stressing about anthropogenic global warming (AGW)?

Plate tectonics and continental drift are continuing processes that lead to new mountain building and submergence of areas, control the climate of the area, and influence sea currents. These changes are minute and take millions of years. Earthquakes and volcanoes are manifestations of and adjustments to these movements of drifting continents.

A geological perspective on climate change refers to the study of past climate changes and their causes and how this information can be used to better understand and predict future climate patterns. It provides valuable insights into past climate changes, their causes, and their potential impacts by using this information to inform future climate projections.

Geological evidence such as sedimentary rocks, ice cores, and tree rings provides insights into climate change over millions of years. Geological evidence shows that climate change has occurred naturally throughout Earth's history, with periods of warming and cooling lasting for hundreds of thousands or even millions of years.

The mean global temperature over geological time has varied significantly, with periods of both warmth and cold. Throughout Earth's history, there have been several major climate events and eras. Here is a broad overview.

Hadean and Archean Eons (4.6 billion to 2.5 billion years ago): During this period, Earth's surface was extremely hot, with molten rocks and a volatile atmosphere. Detailed temperature records are not available at this time.

Proterozoic Eon (2.5 billion to 541 million years ago): The early part of this aeon was marked by several ice ages, including the famous "Snowball Earth" events when the planet's surface was largely covered in ice. However, there were also periods of warmer climates.

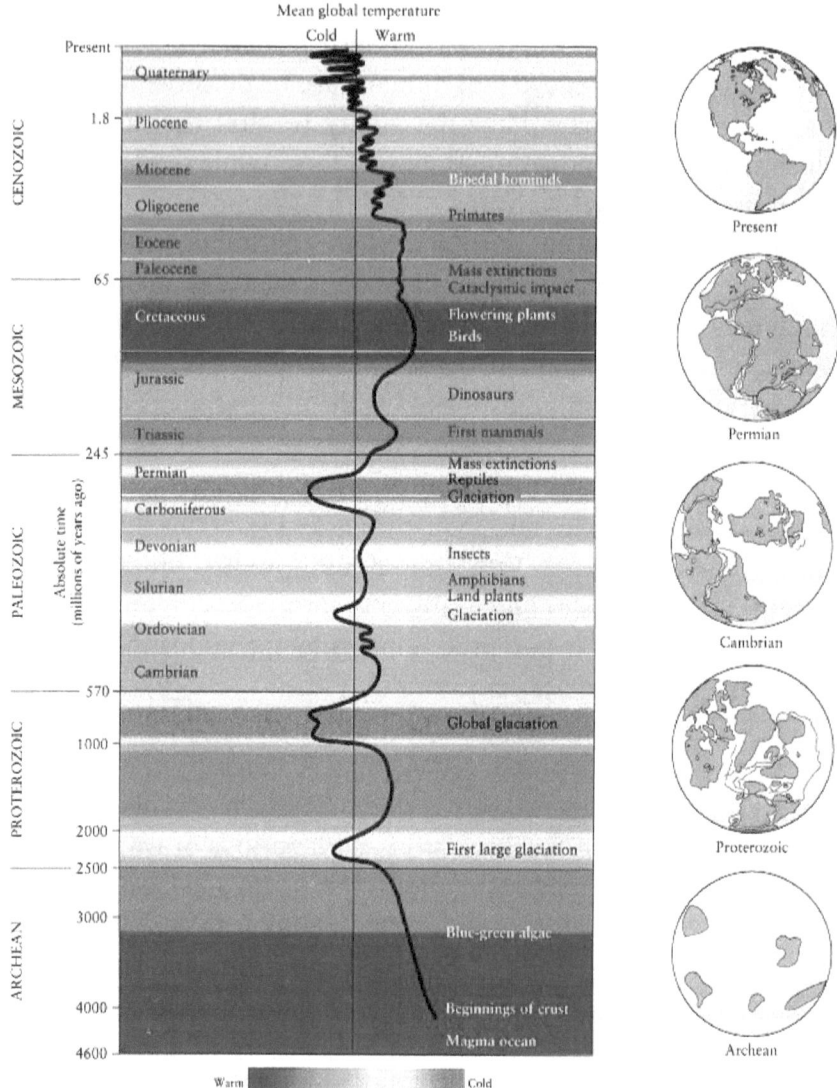

Figure 5: Mean Global temperature

Source: https://cdn.britannica.com/92/192592-004-E8A9C1D0/timeline-developments-climate-change.jpg

Phanerozoic Eon (541 million years ago to present): This aeon is divided into three eras: Palaeozoic, Mesozoic, and Cenozoic.

Palaeozoic Era (541 million to 252 million years ago): The early Palaeozoic was generally warm, with tropical conditions and high sea levels. However, there were some glaciations towards the end of this era.

MePhanerozoic Eon (541 million years ago to present): This aeon is divided into three eras: Paleozoic, Mesozoic, and Cenozoic.

Palaeozoic Era (541 million to 252 million years ago): The early Palaeozoic was generally warm, with tropical conditions and high sea levels. However, there were some glaciations towards the end of this era.

Mesozoic Era (252 million to 66 million years ago): The Mesozoic era is often referred to as the "Age of Dinosaurs". It had a generally warmer climate compared to today, with higher atmospheric carbon dioxide levels. However, there were fluctuations, including periods of cooling and warming.

Cenozoic Era (66 million years ago to present): The Cenozoic era is the current era and can be divided into the Paleogene, Neogene, and Quaternary periods. Overall, this era has experienced a cooling trend, leading to the formation of ice caps in Antarctica and the Arctic. The Quaternary period began around 2.6 million years ago and has seen repeated glacial and interglacial cycles, including the most recent ice age.

Mesozoic Era (252 million to 66 million years ago): The Mesozoic era is often referred to as the "Age of Dinosaurs." It had a generally warmer climate compared to today, with higher atmospheric carbon dioxide levels. However, there were fluctuations, including periods of cooling and warming.

Cenozoic Era (66 million years ago to present): The Cenozoic era is the current era and can be divided into the Paleogene, Neogene, and Quaternary periods. Overall, this era has experienced a cooling trend, leading to the formation of ice caps in Antarctica and the Arctic. The Quaternary period began around 2.6 million years ago and has seen repeated glacial and interglacial cycles, including the most recent ice age.

Phanerozoic Eon (541 million years ago to present): This aeon is divided into three eras: Paleozoic, Mesozoic, and Cenozoic.

Palaeozoic Era (541 million to 252 million years ago): The early Palaeozoic was generally warm, with tropical conditions and high sea levels. However, there were some glaciations towards the end of this era.

Mesozoic Era (252 million to 66 million years ago): The Mesozoic era is often referred to as the "Age of Dinosaurs". It had a generally warmer climate compared to today, with higher atmospheric carbon dioxide levels. However, there were fluctuations, including periods of cooling and warming.

Cenozoic Era (66 million years ago to present): The Cenozoic era is the current era and can be divided into the Paleogene, Neogene, and Quaternary periods. Overall, this era has experienced a cooling trend, leading to the formation of ice caps in Antarctica and the Arctic. The Quaternary period began around 2.6 million years ago and has seen repeated glacial and interglacial cycles, including the most recent ice age.

Geologists also study the natural processes that contribute to climate change, such as volcanic eruptions, changes in the Earth's orbit, and variations in solar radiation. Geological evidence also provides insight into the potential impacts of climate change, such as rising sea levels, changes in precipitation patterns, and the frequency and intensity of extreme weather events. By studying past climate changes, geologists can help inform predictions for future climate scenarios and assist in developing strategies for mitigating and adapting to climate change.

Geological evidence provides a valuable long-term perspective on climate change, enabling scientists to investigate how the Earth's climate has transformed over millions of years. Scientists analyse various geological perspectives to study climate change, including paleoclimate records, plate tectonics, volcanic activity, the carbon cycle, and sea level changes. Paleoclimate records, such as ice cores, sediment layers, and fossil records, provide evidence of past changes in temperature, atmospheric composition, sea level, and other key indicators of climate change. Plate tectonics, the movement of the Earth's crust, has played a significant role in climate change by altering ocean currents and atmospheric circulation. Volcanic activity can release greenhouse gases that warm the planet or particles that cool it. The carbon cycle, the transfer of carbon between the atmosphere, oceans, and Earth's crust, regulates the Earth's climate over geologic time. Sea level changes, influenced by melting ice sheets, shifts in ocean currents, and tectonic activity, can cause the formation of new coastlines, the erosion of shorelines, and modifications in ocean circulation patterns.

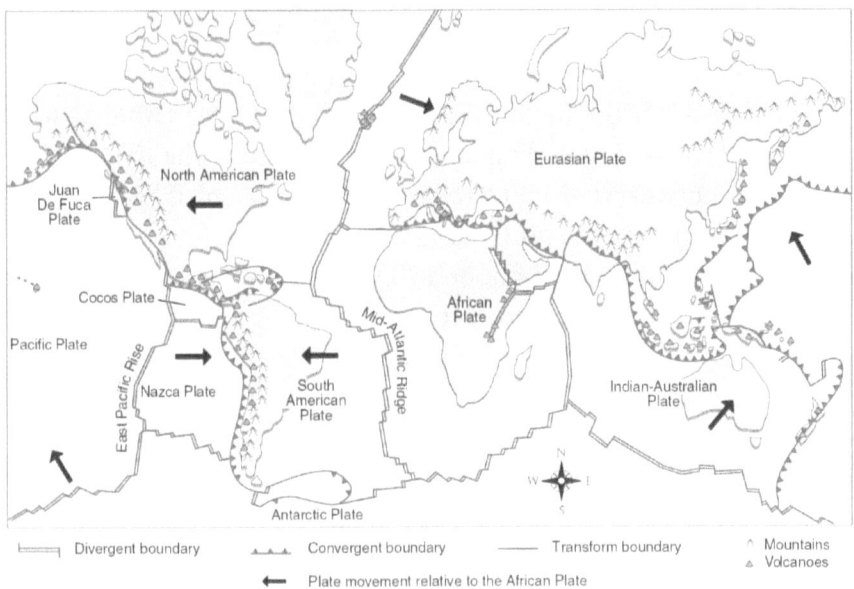

Figure 6: Continental Drift

Source: https://exampariksha.com/wp-content/uploads/2014/12/Continental-drift-21.jpg

By studying paleoclimate records, plate tectonics, volcanic activity, the carbon cycle, and sea level changes, scientists can gain insights into how the Earth's climate has responded to natural changes in the past and use this knowledge to better understand and mitigate the impacts of human-caused climate change in the present and future.

The Himalayas, the highest mountain range in the world, were once a sea by the name of the Tethys Sea. By the way, the Himalayas are still rising because of the pushing of the Indian Plate against the Eurasian Plate. Geological records show that there have been several large variations in the Earth's climate. Numerous natural factors, such as variations in the Earth's orbit, emissions from volcanoes, and carbon dioxide levels, have contributed to these. Global climate change has typically occurred slowly over thousands or millions of years. There have been several major ice ages throughout geological history due to the interplay between life, the carbon cycle, and climate.

Continents near the poles favour the development of ice sheets. The location of plates and the related development of mountain belts in warm and humid climate zones favour enhanced chemical weathering. Changes in the planet's orbit can be attributed to a number of factors, including changes in the tilt of the axis (which occurs roughly every 40,000 years), the precession of the rotation axis (which occurs roughly every 23,000 years), and eccentricity (which follows a cycle of about 100,000 years), affect the amount of energy that the Earth's surface can absorb. Additionally, changes in ocean circulation also contribute to this phenomenon. NASA has developed an elaborate animated model that illustrates the complex dynamics of ocean currents and the way they interact with winds on the surface. The video from NASA provides a concise overview of multiple satellite systems, presenting interconnected graphics.

The global climate is complex, as there are many feedback mechanisms (both positive and negative). Some parameters amplify others (for example, Greenhouse gases can increase after warming is initiated by changes in the Earth's orbit and amplify the resulting global warming). The study of the Earth's geological history depends on understanding its climate, and it is well known that a variety of factors, including positive and negative feedback mechanisms that can change certain parameters over time, influence the global climate.

By examining the geologic time scale, which categorises Earth's history into smaller time segments, researchers can gain insight into the intricate interplay between the global climate and various factors, including positive and negative feedback mechanisms that amplify or diminish certain parameters over time. Earth's history is divided into a hierarchical series of smaller chunks of time, referred to as the geologic time scale. These divisions, in descending length of time, are called aeons (encompassing billions of years), eras (typically characterised by a distinctive fossil record), periods (characterised by distinctive rock units), epochs, and ages (Figure 7).

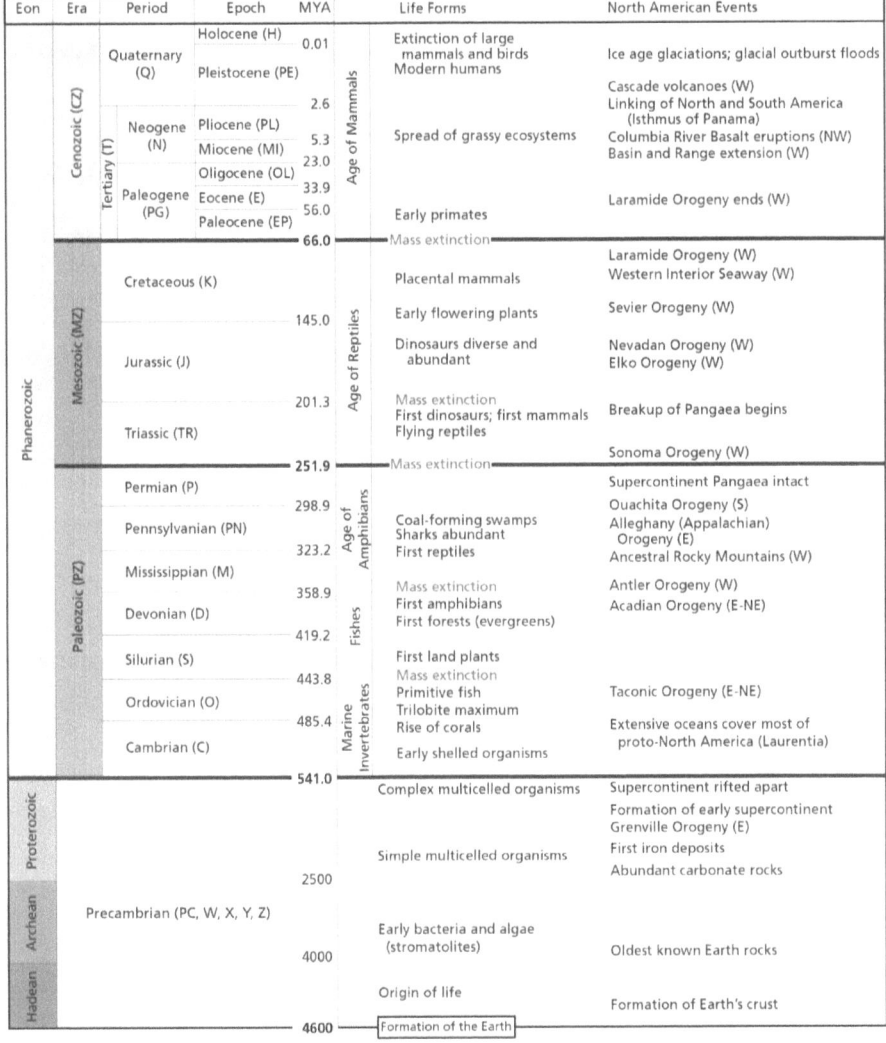

Figure 7: Geological Time Scale

These units are classified based on Earth's rock layers, or strata, and the fossils found within them. From examining these fossils, scientists know that certain organisms are characteristic of certain parts of the geologic record. The study of this correlation is called stratigraphy. The last and most recent period of the geological time scale is the Holocene Epoch. It began 11,700 years ago, after the last major ice age.

The Anthropocene Epoch is an unofficial unit of geologic time used to describe the most recent period in Earth's history when human activity started to have a significant impact on the planet's climate and ecosystems. In 2000, biologists Eugene Stormer and Paul Crutzen coined the term "anthropocene", which is derived from the Greek words "anthropo" for "man" and "cene" for "new".

The International Union of Geological Sciences (IUGS), the global organisation that names and defines epochs, has not officially adopted the term "Anthropocene", and scientists continue to debate whether it differs from the Holocene in terms of both the term itself and the phenomenon it encompasses. The primary question that the IUGS needs to answer before declaring the Anthropocene an epoch is whether humans have changed the Earth system to the point that it is reflected in the rock strata.

The next issue was when it started, which has also been hotly contested among scientists who believe the Anthropocene describes a new geological period. A popular theory is that it began at the start of the Industrial Revolution of the 1800s when human activity had a great impact on carbon and methane in the Earth's atmosphere. Others think that the beginning of the Anthropocene should be in 1945. This is when humans tested the first atomic bomb and then dropped atomic bombs on Hiroshima and Nagasaki, Japan. The resulting radioactive particles were detected in soil samples globally.

In 2016, the Anthropocene Working Group agreed that the Anthropocene is different from the Holocene and began in the year 1950, when the Great Acceleration, a dramatic increase in human activity affecting the planet, took off.

The geological record shows that carbon dioxide decreased during much of the Phanerozoic Aeon, leading to a widespread drop in global temperature (Alejandro Cearreta, 2022). During the Cretaceous period, carbon dioxide increased three to six times higher than pre-industrial levels, with sea levels more than 60 m above present levels. The Quaternary period is characterised by geological time units linked to climate events. Numerous warm interglacial phases caused by changes in the Earth's orbit and axis of rotation interrupted the Pleistocene Ice Age. The last of these warm phases is the current Holocene Epoch when the climate stabilised. The three parts of the Holocene—the Greenlandian, the Norgrippian, and the Meghalayan—are based on geochemical markers found in ice cores. These markers show that the climate changed quickly on a global scale 11,700, 8,200, and 4,200 years ago, respectively. The human species, Homo sapiens, emerged on the planet about 300,000 years ago.

The magnitude of the environmental changes that took place in the last two centuries is vastly different from the large-scale and synchronous changes that took place in the past (Syvitski, J., et al., 2020). The presence of increased greenhouse gases in ice core surveys since the late 18th century has brought about a new geological time unit, the Anthropocene, in Earth's history (Crutzen, P. J., & Stoermer, E. F., 2000). The Anthropocene is the only chronostratigraphic geological time unit contained within recorded and documented human history.

Earth's climate has been cooling for the past 50 million years, from 6 to 7^0 degrees Celsius above today's global average temperature. The cooling led to the formation of the caps in Antarctica 34 million years ago and in the northern hemisphere around 2.6 million years ago. The decrease in atmospheric carbon dioxide was the primary cause of this cooling (Summerhayes, Colin P., et al. 2013).

The evidence of climate change is preserved in many geological settings, like marine or lake sediments, ice sheets, fossil corals, stalagmites, and fossilised tree rings. The ice sheet core drills indicate a record of polar temperatures and, thereby, atmospheric conditions going back 120,000 years in Greenland and 800,000 years in Antarctica. Similarly, the ocean sediments preserve records reaching tens of millions of years, and the older sedimentary rocks even go back hundreds of millions of years.

Climate change is a natural geological and meteorological phenomenon like natural warming and cooling, earthquakes, volcanic eruptions, tsunamis, and floods that cause adverse societal hazards. By being prepared to handle them, mitigate their effects, and manage the damage they cause, they should be handled in the same way we handle these hazards after they occur. Attempting to stop climate change by reducing human-induced carbon dioxide is utterly futile. Rational climate policies must be based on adaptations to change as and when it occurs (Carter, R.M., 2007).

5.3 The Relationship between Carbon Dioxide and the Earth's Atmosphere

Carbon is an essential component of life on our planet (Figure 8). All living tissue, including that of plants, animals, and humans, is composed of carbon-containing compounds. Carbon can also be found in wood, coal, marble, and limestone, as well as petroleum-based plastics and fuels. The binding capacity of carbon atoms is responsible for the variety of forms. Experts now know far more than a million different carbon compounds and new ones are discovered every year, forming a separate discipline in chemistry.

Carbon is naturally absorbed, released, chemically bound, or otherwise transformed at any given time anywhere on the planet due to its chemical properties and widespread distribution. This means that carbon is always in motion and moves through all components of the Earth's system over time. Carbon takes a different amount of time for each step of this journey. Carbon or its compounds are sometimes released or absorbed in a matter of minutes (respiration, combustion), while other times, they are stored for thousands or millions of years (photosynthesis, dissolution in seawater, permafrost, formation of fossil raw materials). Various biogeochemical processes on land and in the ocean control the carbon dioxide concentration in the Earth's atmosphere, which is critical for climate. The greenhouse gas is either removed from the atmosphere and stored (carbon sinks) or released during these processes (carbon sources).

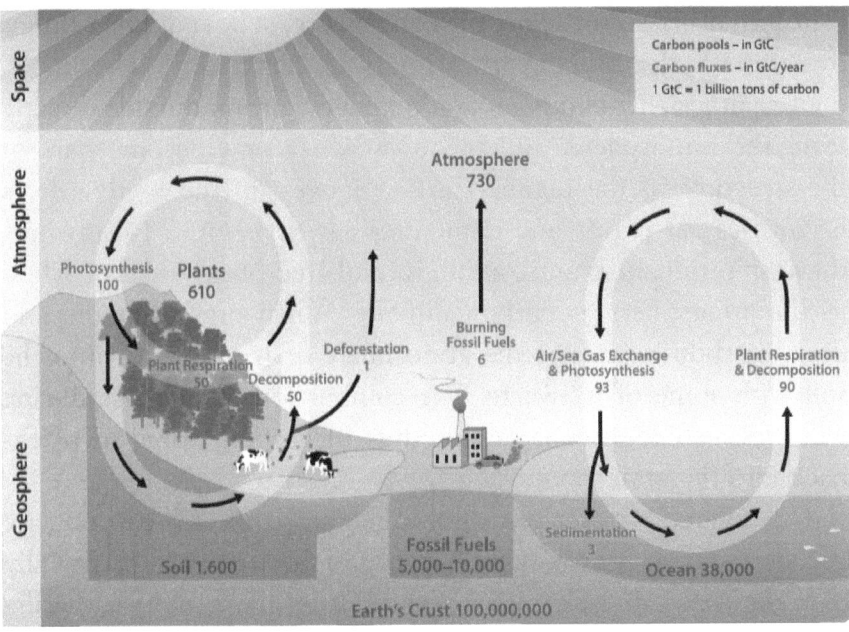

Figure 8: Carbon Cycle
Source: https://eunit.plt.org/wp-content/uploads/sites/2/2017/11/6-8_Global CarbonCycle_Image.png

While most of the Earth's carbon can be found in the geosphere, carbon is found in all living things, including soils, the ocean, and the atmosphere. Carbon is the primary building block of life, including DNA, proteins, sugars, and fats. One of the most important carbon compounds in the atmosphere is carbon dioxide, while in rocks, carbon is a major component of limestone, coal, oil, gas, and even diamond. Carbon cycles through the atmosphere, biosphere, geosphere, and hydrosphere via processes that include photosynthesis, fire, the burning of fossil fuels, weathering, and volcanism.

By understanding how human activities have altered the carbon cycle, we can explain many of the climate and ecosystem changes we are experiencing today and why this rapid rate of change is largely unprecedented in the Earth's history. Carbon is transferred between the ocean, atmosphere, soil, and living things over time scales ranging from hours to centuries. For example, photosynthesizing plants on land removes carbon dioxide directly from the atmosphere, and those carbon atoms become part of the structure of the plants. Carbon moves up the food web as herbivores eat plants and carnivores eat herbivores. Meanwhile, the respiration of plants, animals, and microbes returns carbon to the atmosphere as carbon dioxide. When organisms die and decay, carbon returns to the atmosphere or is integrated into the soil with some of its waste. The combustion of biomass during wildfires also releases large amounts of carbon stored in plants back into the atmosphere.

On longer timescales, significant amounts of carbon are transferred between rocks, the ocean, and the atmosphere, typically over thousands to millions of years. For example, the weathering of rocks removes carbon dioxide from the atmosphere. The resulting sediments, along with organic material, can be eroded and transported from the land to enter the ocean, where they sink to the bottom. Algae, plants, and animals can incorporate this land-based carbon as well as carbon atoms in carbon dioxide that the ocean absorbs from the atmosphere into calcium carbonate ($CaCO3$) shells. These shells become buried. As the successive layers of sediment are compressed and cemented, they are turned into limestone. Over millions of years, these carbon-bearing rocks can be exposed to sufficient heat and pressure to melt, causing them to release their carbon back into the atmosphere as carbon dioxide via volcanism. Some of these rocks will also be exposed at the surface of the earth through mountain building and weathering, and the cycle begins again. Carbon from the mantle is also released into the atmosphere as carbon dioxide through volcanic activity.

Carbon is also transferred to rocks from the biosphere via the formation of fossil fuels, which form over millions of years. Fossil fuels are derived from the burial of photosynthetic organisms, including plants on land (that primarily form coal) and plankton in the oceans (that primarily form oil and natural gas). While buried, this carbon is removed from the carbon cycle for millions to hundreds of millions of years.

Human activity, especially the burning of fossil fuels, has dramatically increased the exchange of carbon from the ground back into the atmosphere and oceans. This return of carbon back into the atmosphere as carbon dioxide is released will occur at a rate that is hundreds to thousands of times faster than it took to bury it and much faster than it can be removed by the carbon cycle (for example, by weathering). Thus, the carbon dioxide released from burning fossil fuels accumulates in the atmosphere, increasing average temperatures through the greenhouse effect and dissolving in the ocean, causing ocean acidification.

The Global Carbon Project estimates that human activities will generate around 33.6 billion metric tonnes of carbon dioxide emissions in 2021. This includes emissions from burning fossil fuels (such as coal, oil, and gas), cement production, and deforestation. To arrive at the figure of 33.6 billion metric tonnes of total carbon dioxide emissions from human activity in 2021, the GCP uses a combination of methods, including bottom-up and top-down approaches.

Bottom-up approaches involve estimating emissions from individual sources, such as power plants and transportation, and then summing these estimates to obtain a total for a particular region or country. Top-down approaches involve measuring the concentration of greenhouse gases in the atmosphere and using mathematical models to estimate the sources and sinks of these gases.

Bottom-up approaches rely on data from individual sources of emissions, such as power plants, factories, and transportation, and then aggregate these estimates to estimate emissions for a particular region or country. The accuracy of bottom-up approaches can depend on the quality and availability of data, which can vary significantly across countries and sectors. In some cases, emissions data may be incomplete or inaccurate, which can lead to an underestimation or overestimation of emissions.

According to the Global Carbon Project, the terrestrial biosphere (land plants and soils) currently absorbs about 30% of the CO_2 emissions from human activities each year. This is sometimes referred to as the "land carbon sink". However, the effectiveness of this sink can change from year to year and depends on a number of variables, such as the climate, changes in land use, and disturbances like fires and insect outbreaks.

The GCP uses a method to calculate the net CO_2 flux between the atmosphere and the land surface, which takes into account both CO_2 emissions from human activity and CO_2 absorption by land ecosystems. The net CO_2 flux can be estimated using a combination of atmospheric measurements and modelling, which provide information on the amount of CO_2 in the atmosphere, the sources and sinks of CO_2, and the transport of CO_2 between the atmosphere and the land surface.

The GCP also uses an estimation method for the CO_2 uptake by land ecosystems, which refers to the amount of CO_2 plants and soils absorb through photosynthesis and other processes. This can be estimated using a combination of satellite data, ground-based measurements, and modelling.

By combining these different methods, the GCP estimates that the terrestrial biosphere currently absorbs about 30% of the CO_2 emissions from human activities each year.

We may be tempted to believe that if we triple the plantation, the plants will absorb the rest of the carbon dioxide produced by human activity. However, it is highly unlikely that tripling the plantation would completely offset the carbon dioxide that human activity produces. Climate variability, land use change, and disturbances like fires and insect outbreaks can all have an impact on the effectiveness of the land carbon sink, which refers to the capacity of plants and soils to absorb and store carbon.

Even if it were possible to plant enough trees to completely offset human CO2 emissions, there are other factors to consider. For example, increasing the amount of land dedicated to plantations could impact biodiversity, water resources, and local communities. Moreover, planting trees is not a permanent solution to the problem of CO2 emissions, as carbon stored in trees can be released back into the atmosphere through natural processes such as respiration and decomposition.

The United Nations Population Division projects that the world population will reach 9.7 billion in 2050 and peak at around 11 billion in the second half of the 21st century before gradually declining. How will it affect the emission of carbon dioxide? It is possible that the emission of carbon dioxide could be reduced proportionately as well. A declining population could lead to reduced energy consumption, as fewer people would need energy for transportation, housing, and other daily activities. This could lead to lower carbon emissions from fossil fuels, a major contributor to greenhouse gas emissions.

The rate of exchange and the distribution of carbon in the Earth's system is affected by various human activities and environmental phenomena, including the burning of fossil fuels, which rapidly releases carbon dioxide, a greenhouse gas, into the atmosphere, increases average global temperatures and causes ocean acidification.

Agricultural activities release carbon dioxide and methane (CH_4, greenhouse gases) into the atmosphere. For example, methane is produced from the digestion of plant material by cattle and from the bacteria that thrive in rice fields. Carbon dioxide is released from the burning of fossil fuels to power farming equipment, from the mining of minerals, and from the making of fertiliser. The growing of crops and the raising of livestock also affect local productivity, biomass, and rates of photosynthesis, respiration, and decay of organic material.

Deforestation lowers photosynthesis rates and, as a result, the amount of carbon dioxide that plants can absorb. Trees absorb carbon dioxide from the atmosphere and store it in their wood, leaves, bark, and roots as they grow. The carbon is returned to the atmosphere when downed trees are left to rot or if the trees are intentionally set on fire, which is a common means of deforestation. Thus, deforestation typically releases carbon dioxide unless all the material is used for construction or paper products.

The amount of soil that remains frozen throughout the year, known as permafrost, holds methane, a greenhouse gas. As long as temperatures remain consistently cold, any organic matter present in the permafrost decays at a very slow pace and stays trapped in the soil. However, as global temperatures rise, the permafrost is melting and releasing methane into the atmosphere. The higher temperatures accelerate the rate of decay and increase the release of greenhouse gases into the atmosphere, which further amplifies this process.

Over the course of millions of years, the rate of sedimentation and the burial rate of organic matter has varied, leading to changes in the amount of carbon available for decay and the amount stored in rocks. For instance, when dead plants and plankton are buried at a higher rate, the rate of decay decreases, resulting in a faster formation of fossil fuels.

Over millions of years, processes in the rock cycle can change carbon dioxide levels in the atmosphere. For example, the metamorphic reactions that occur under heat and pressure can release carbon dioxide. In contrast, the weathering of rocks that occurs when carbon dioxide dissolves into rainwater to form carbonic acid (H_2CO_3) reduces the amount of carbon dioxide in the atmosphere. Warming can increase these weathering reactions, but not at a rate that can offset the increase in carbon dioxide due to human activities.

Plate tectonics-driven changes in the rate of volcanism can significantly alter the amount of carbon dioxide in the atmosphere, but they do so over millions of years on a much longer timescale than human timescales.

A review of the research papers and literature concerning the environmental consequences of increased levels of atmospheric carbon dioxide leads to the conclusion that increases during the 20th and early 21st centuries have produced no deleterious effects on Earth's weather and climate. Increased carbon dioxide has, however, markedly increased plant growth. Predictions of harmful climatic effects due to future increases in hydrocarbon use and minor greenhouse gases like carbon dioxide do not conform to current experimental knowledge (Soon, Willie, et al., 1999).

Carbon dioxide, a colourless gas, has been present in the atmosphere in amounts ranging from a few hundred to a few thousand parts per million (ppm). The average values of carbon dioxide over the past hundred thousand years, as inferred from the ice core data, have been found to be 180 ppm during glacial periods and 280 ppm during interglacial periods (Petit, J. et al., 1999). These values are 30–50 per cent lower than the original atmospheric values (Anon, 2000; Jaworowski, Z., 2007). Between 444 and 353 million years ago (Phanerozoic), there was extensive glaciation, and the atmospheric carbon dioxide level was 17 times higher than it is today (Chumakov, N.M., 2004).

As of 2021, the concentration of carbon dioxide (CO_2) in the Earth's atmosphere is approximately 416 parts per million (ppm) by volume. This means that for every one million molecules of air, about 416 of them are CO_2 molecules. To express this as a percentage, we can divide the concentration of CO_2 by the total concentration of all gases in the atmosphere, which is around 99.96%. Thus, 416 ppm of CO2 out of 999,584 ppm of total gases comes out to be just 0.0416%. Therefore, carbon dioxide currently makes up about 0.0416% of the Earth's atmosphere by volume. It is estimated that about 63% of the total carbon dioxide (CO2) in the atmosphere is of anthropogenic (human-caused) origin, primarily from burning fossil fuels, land-use changes, and other human activities. To calculate 63% of 0.0416, we can multiply 0.0416 by 0.63, which is 0.0416 x 0.63 = 0.026208. Therefore, approximately 0.026208% of the Earth's atmosphere by volume is anthropogenic carbon dioxide.

The levels of carbon dioxide in the recent past were very low, going by the earlier geological history. Carbon dioxide has declined from 1000 ppm in the Cenozoic (60 million years ago) (Lowenstein, T. K., & Demicco, R. V., 2006). Evidently, there is nothing unusual or dangerous about the current carbon dioxide content in the atmosphere. On the contrary, increasing carbon dioxide in the range of 200–1000 ppm has been seen to be beneficial for plant growth as it increases the plant's efficiency of water use (Eamus, D., 1996), (J.-G., et al., 2007) and (Saxe, H., Ellsworth), (D., & Heath, J., 1998).

Evidently, given the above facts, there is no cause for concern to assume that atmospheric carbon dioxide between 500 and 1000 ppm would be dangerous and have adverse ecological effects.

About 63% of the current atmospheric carbon dioxide (CO2) concentration of 415 parts per million (ppm) is estimated to be the result of human activity, primarily from burning fossil fuels like coal, oil, and gas, as well as deforestation and other land-use changes. This estimate is based on measurements of the isotopic composition of atmospheric CO2, which allows scientists to distinguish between carbon from fossil fuels and carbon from natural sources like volcanoes and the ocean.

Carbon in fossil fuels is relatively depleted in the heavy isotope carbon-13 (13C) compared to the lighter isotope carbon-12 (12C).

A large body of research supports the claim that carbon in fossil fuels is relatively depleted in the heavy isotope carbon-13 (13C) compared to the lighter isotope carbon-12 (12C).

Meyers, P. A., and Lallier-Vergès, E. (1999) discuss the isotopic composition of carbon in sedimentary organic matter from different types of lakes around the world, including lakes that have received inputs of organic matter from terrestrial plants and lakes that have received inputs of organic matter from aquatic plants. The authors find that the carbon in organic matter from terrestrial plants, including those that contributed to the formation of fossil fuels, has a lower 13C/12C ratio compared to the carbon in organic matter from aquatic plants.

Lee and Fung (Lee, X., & Fung, I., 2008) discuss the isotopic composition of carbon in sediment cores from Lake Balkhash in central Asia. The authors find that the carbon in sedimentary organic matter from the period of fossil fuel burning (mid-20th century to present) has a lower 13C/12C ratio compared to the carbon in sedimentary organic matter from earlier periods.

Schimmelmann et al. (2006) discuss the isotopic composition of carbon in coalbed methane from various coal deposits around the world. The authors find that the carbon in coalbed methane has a lower 13C/12C ratio compared to the carbon in methane from other sources, such as natural gas from conventional reservoirs, which is consistent with the isotopic signature of carbon in fossil fuels.

The isotopic ratio of carbon dioxide (CO_2) in the atmosphere is roughly 1 part 13C for every 99 parts 12C, which means that 13C constitutes around 1% of the total carbon in the atmosphere. This ratio is referred to as 13C and is usually expressed in parts per thousand. It is calculated as the difference between the ratio of 13C to 12C in a given sample and the ratio in a standard.

However, the ratio of 13C to 12C can vary depending on the source of the CO_2. For instance, when compared to CO_2 from natural sources like respiration, photosynthesis, and volcanic activity, CO_2 from the burning of fossil fuels is relatively depleted at 13C. This is because plants preferentially take up the lighter isotope, 12C, during photosynthesis, and fossil fuels are made up of organic matter that has been buried for millions of years, during which time some of the 13C has decayed into 14C and been lost from the system. As a result, the 13C value of CO_2 produced by burning fossil fuels is usually around -27, whereas the 13C value of CO_2 produced by natural sources is usually between -8 and -10.

The isotopic composition of carbon in the atmosphere is an important tool for studying the global carbon cycle and the sources and sinks of atmospheric CO_2. Scientists use measurements of atmospheric CO_2's 13C content to track changes in the carbon cycle and determine the relative contributions of various CO_2 emissions from fossil fuel combustion, shifting land use, and natural processes.

By releasing significant amounts of carbon dioxide into the atmosphere before it can be absorbed by natural sinks, such as plants and oceans, human activities are upsetting the equilibrium of the carbon cycle. This leads to an accumulation of carbon dioxide in the atmosphere, which contributes to global warming.

It is true that human activities are contributing to global warming and other environmental challenges that could have negative impacts on the Earth's habitability. It is also worth noting that humanity has faced existential threats before, including pandemics, wars, and other disasters. While the challenges we face today are certainly significant, it is important to remember that humans have shown remarkable resilience and adaptability in the face of adversity throughout history.

Although human activities are causing global warming and other environmental issues that could potentially harm the habitability of the Earth, it is my belief that we are not facing an existential threat from climate change. Hence, if one believes that the current human-caused global warming is not severe enough to pose a significant danger to humanity, one might wonder why so much concern and attention is given to this issue.

While it is not possible to completely eliminate the impacts of global warming and climate change, it is possible to mitigate their severity and reduce their potential impacts through a combination of actions, including reducing greenhouse gas emissions, promoting renewable energy, protecting natural ecosystems, and developing sustainable technologies and practices.

It is true that electricity is a versatile and critical form of energy that powers much of modern society, regardless of the energy source used to generate it. In many cases, electricity can be used to power devices, buildings, and transportation more efficiently than other forms of energy, making it a valuable component of a sustainable energy system.

Since plants need carbon dioxide to produce carbohydrates and other organic compounds during the process of photosynthesis, it is true that carbon dioxide is a necessary component of the food chain. However, too much carbon dioxide in the atmosphere can be harmful to human existence, primarily due to its role in causing climate change.

When greenhouse gases, such as carbon dioxide and methane, are emitted into the atmosphere, they trap solar heat, resulting in an increase in the Earth's temperature. This can have various negative effects on human existence, including:

1. Extreme weather conditions: The warming of the Earth's atmosphere can lead to more frequent and severe weather events like hurricanes, floods, and heat waves. These events can cause substantial damage to infrastructure, homes, and communities.
2. Rising sea levels: The melting of polar ice caps due to increased temperatures causes sea levels to rise, resulting in flooding in coastal areas and contamination of freshwater resources with salt water.
3. Health impacts: Rising temperatures can directly affect human health, resulting in higher rates of heatstroke, dehydration, and respiratory illnesses due to increased air pollution.

Therefore, while carbon dioxide is an essential component of the food chain, too much of it in the atmosphere can have negative consequences for human existence.

Before the Industrial Revolution, the amount of carbon dioxide in the atmosphere was relatively stable because the carbon cycle was generally in balance. This means that the amount of carbon dioxide released into the atmosphere from natural processes (such as respiration, volcanic activity, and wildfires) was roughly balanced by the amount of carbon dioxide removed from the atmosphere through natural processes (such as photosynthesis and absorption by the ocean).

However, with the onset of industrialisation and the burning of fossil fuels, the amount of carbon dioxide released into the atmosphere has increased dramatically. This has disrupted the natural balance of the carbon cycle and caused the amount of carbon dioxide in the atmosphere to rise significantly.

In terms of the percentage of carbon dioxide produced by human activity that is offset by plants and vegetation, it is difficult to say exactly what this percentage was before the Industrial Revolution. However, it is likely that the natural carbon cycle was able to absorb a much larger percentage of carbon dioxide emissions before the rise of industrialisation and the burning of fossil fuels.

According to current estimates, the ocean and plants only absorb about 30% of human-caused carbon dioxide emissions, with the remaining 70% staying in the atmosphere and causing climate change. This has led to a significant increase in atmospheric carbon dioxide concentrations, which are now at their highest levels in at least 800,000 years.

Suppose only 30% of the carbon dioxide produced by human activity is offset by vegetation and plantations, and the ratio of carbon dioxide offset and produced is gradually and continuously decreasing. Does that mean we will reach a point when zero carbon dioxide is offset?

It is possible that the ability of plants and vegetation to absorb carbon dioxide may be limited in the future, especially if carbon dioxide emissions continue to increase. Climate change can also affect the ability of plants to absorb carbon dioxide, as changes in temperature, precipitation, and other factors can alter the rate of photosynthesis and the overall health of ecosystems.

However, it is important to note that the exact threshold at which plants and vegetation will no longer be able to offset carbon dioxide emissions is difficult to predict, as it depends on a range of factors, including future emissions levels, climate change impacts, and land-use changes. It is also possible that technological innovations or changes in human behaviour could help reduce carbon dioxide emissions and mitigate the effects of climate change.

CHAPTER 6

The Business

Among the many theories of climate change, why has the spotlight been focused solely on the anthropogenic theory? Carbon markets have emerged as important policy instruments for battling climate change because this is the only theory that offers the chance to stir up fear, make money, and enlist market mechanisms for climate mitigation on behalf of corporations. In his book "Heaven on Earth: Global Warming, the Missing Science" (Ian Plimer, 2009), Australian Earth scientist Ian Plimer asserted that global warming is a lie and that the entire international climate science, politics, and media have come together to support this deception.

Al Gore, the primary advocate and champion of the anthropogenic global warming (AGW) theory, is believed to have commercial motivations at the heart of his involvement. He directed the documentary "An Inconvenient Truth" and became a prominent supporter of AGW. He founded his own eco-friendly business, Generation Investment Management, and was a partner at Kleiner Perkins, a venture capital firm, from 2007 to 2016. Additionally, he co-founded a cable television network in 2005, later sold to Al Jazeera in 2013. It seems that his interests are primarily financial, and he appears to be indifferent to the issue of climate change.

A narrative has emerged that justifies and secures corporate interests in climate policy. Climate policies are based on the common assumption that we can achieve economic growth without causing significant harm to the environment (Wright, Christopher, and Daniel Nyberg, 2015). The stories that corporations construct about climate change are not merely a means to conceal their environmental damage but also a way to profit from it by utilising a new approach known as carbon trading and attempting to transform the broader political and social environment in which they operate. This entails ensuring that the notion of expanding markets and promoting economic growth is perceived as the correct and sole solution to all of humanity's problems.

To sustain this falsehood, corporations must actively engage in political lobbying and charitable endeavours to guarantee the continued credibility of the market-driven narrative. The corporate story is particularly persuasive because it employs a cost-benefit analysis, which implies that the threat of climate change can be minimised through a painless exchange between equal priorities, each measured in terms of a particular economic value.

6.1 Carbon Trading-Commodification of a catastrophe

In the 1960s, economist Thomas Crocker and ecologist John H. Dales introduced the idea of carbon trading. They suggested that a market-driven approach to decreasing pollution might be more efficient and successful than conventional command-and-control regulations. This idea was subsequently refined and incorporated into the Kyoto Protocol, which created a structure for nations to exchange carbon emission credits in 1997.

The foundational paper on carbon trading by Crocker, titled "The Structuring of Atmospheric Pollution Control Systems" (Crocker, T. D., 1966), was initially published in 1966 in the journal Essays on Pollution Abatement and Economics. In 1968, Dales and William published a paper titled "Pollution, Property, and Prices" in the Canadian Journal of Economics and Political Science, which further contributed to the concept of carbon trading (Willis, J.R., and Dales, J.H., 1969). These papers by Crocker, Dales, and William established the theoretical basis for carbon trading as a market-oriented strategy for decreasing pollution. This approach gained prominence as a policy instrument for combating climate change in the 1990s and was adopted as part of the Kyoto Protocol in 1997, allowing countries to use it as a mechanism for achieving their emission reduction objectives.

Carbon trading refers to a mechanism that transforms environmental degradation and pollution into a tradeable asset. The system can create a situation known as a "moral hazard," where companies may view it as a chance to persist in their polluting practices by purchasing "emissions credits" to counterbalance their emissions. Advocates of carbon trading, however, contend that it creates a market-driven incentive for businesses to lower their carbon emissions by permitting them to sell extra emissions credits to other companies that have exceeded their allotted emissions levels. Consequently, businesses may have a monetary motivation to invest in cleaner technologies and decrease their carbon footprint.

The commercialisation of carbon emissions and the formation of a carbon trading marketplace provide benefits to an array of stakeholders, such as corporations, governments, and investors. However, the community and consumers are typically excluded from these advantages. Carbon trading can offer a financial incentive for companies to lower their greenhouse gas emissions by providing them with carbon credits when they do so. These credits can be sold to other companies or governments that require offsets for their own emissions. This results in the establishment of a carbon credit market that can create a fresh income source for businesses.

Governments can use carbon trading as a cost-effective approach to meet emissions reduction targets. Rather than enforcing stringent regulations on businesses, governments can set a limit on total emissions and allow companies to exchange carbon credits. This can help lower the overall cost of emissions reductions for both the government and businesses. Carbon trading can also be an opportunity for investors to participate in the growing market for sustainable investments, which are becoming increasingly important in today's world. Investing in companies that are reducing their carbon footprint and sustainable technologies, investors can help support the transition to a low-carbon economy while potentially earning a financial return.

Additionally, carbon trading can also lead to the displacement of communities and the loss of livelihoods, particularly in developing countries where carbon offset projects are often located. These projects may involve the acquisition of land or resources, which can lead to conflicts with local communities and disrupt traditional ways of life. Additionally, some critics have questioned the effectiveness of carbon trading in reducing emissions, claiming that it may just allow businesses to buy their way out of their responsibility to reduce their emissions. Finally, the complexity and opacity of carbon trading markets can create opportunities for fraud and abuse, which can undermine the credibility of the system and erode public trust.

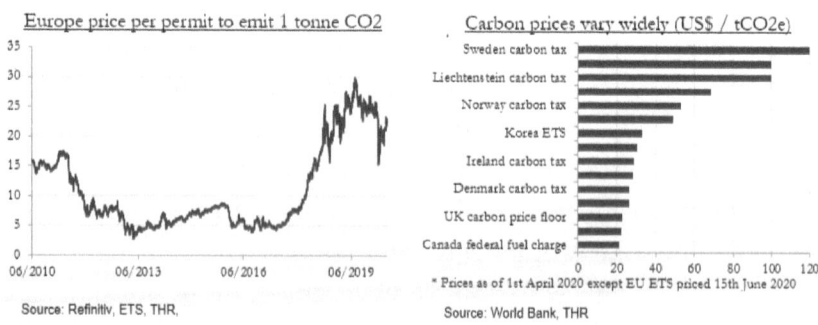

Figure 9: The Rise of Carbon Pricing and Impact on Earnings

Source: https://static.seekingalpha.com/uploads/2020/6/27/51352134-15932451292732494_origin.png

The following is an overview of carbon and emission trading:

The commodification of carbon emissions takes place in two steps: carbon pricing and carbon trading. Carbon trading entails establishing a market for the purchase and sale of permits allowing businesses to emit a certain quantity of greenhouse gases.

A government or international organisation establishes a cap on the total quantity of greenhouse gas emissions companies in a particular sector or jurisdiction are permitted to emit. The companies are then granted permits, or allowances, to emit a certain quantity of greenhouse gases. The total number of permits equals the total emission limit.

Companies that emit less than their allotted allowance can sell their unused excess permits to companies that emit more. This makes the emission or training commodified and establishes a market for permits (Figure 10), which can be purchased and sold as with any other commodity.

At the conclusion of a compliance period (e.g., annually or every few years), businesses must turn in sufficient permits to cover their emissions. If a company emits more than its allotment, it must either purchase additional permits on the market or incur penalties. Governments and regulatory agencies monitor emissions and ensure compliance with the system of emissions caps and permits.

Carbon pricing entails putting a price on greenhouse gas emissions (Figure 9), typically through a tax or levy. Carbon pricing can take various forms, but the fundamental concept is to price greenhouse gas emissions to reflect their social and environmental costs.

A carbon tax is a type of direct tax on greenhouse gas emissions that charges individuals and organisations for each unit of CO_2 or other emission they produce. This tax aims to discourage carbon emissions by making polluters pay for the social and environmental costs of their actions. Clean energy initiatives and other environmental projects can be financed with tax revenue.

The process of modifying carbon, which makes it measurable, exchangeable, and comparable, creates a framework for businesses to assess the challenges posed by climate change. These challenges are categorised into physical risk, regulatory risk, market risk, and reputational risk. However, this approach also reduces complex and interdependent ecological processes to individual and tradable chemical components, which oversimplifies the issue (Wright, Christopher, and Daniel Nyberg, 2015).

During the 1980s, corporations formed the Global Climate Coalition (GCC) with the intention of preventing regulatory initiatives on carbon emissions in the USA, and this was one of the first instances of corporations attempting to shape the political agenda on climate change. The impact of the GCC's actions has been significant. Although moderate corporate voices emerged in the mid-2000s, the financial crisis of 2008 resulted in a resurgence of climate change denial and indifference that influenced the political discourse in many Western democracies. During this period, corporations like ExxonMobil played a pivotal role in shaping climate policy in the US, while mining and fossil fuel interests dominated the shaping of Australian climate policy. Corporate influence has expanded to include NGOs, civil society organisations, and conservation groups, all with the aim of shaping the debate and giving corporations a say in deciding what kind of future is feasible.

These narratives create a particular way of thinking that restricts citizens to the role of passive consumers concerned only with environmental issues. By promoting the idea that purchasing environmentally friendly products is the only meaningful action, consumers and individuals can alleviate their climate-related anxieties. The message conveyed is that one's only responsibility is to consume green products and not to engage in broader political action to address the root causes of climate change.

It is a good idea to innovate and develop alternative energy sources such as solar, wind, solar and tidal, geothermal, etc., not because of climate change but because fossil fuels are limited in supply, will not last forever, and energy demand will increase.

Even if human activity is the cause of climate change, there is no way we can influence its course through taxation, according to the arguments of alarmists.

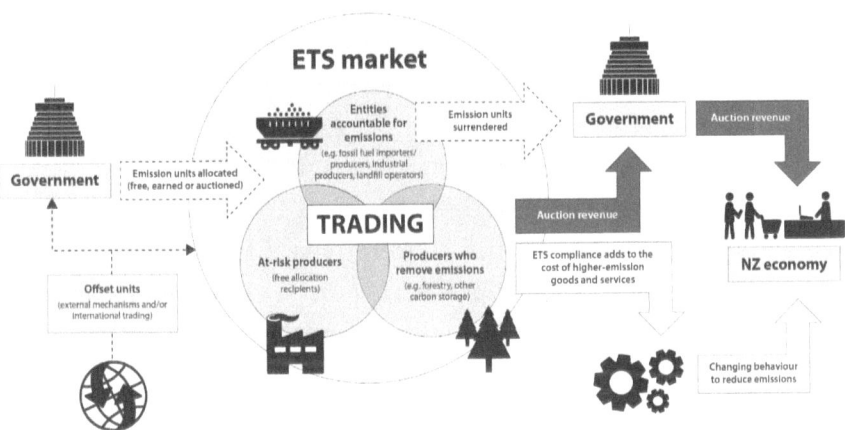

Figure 10: ETS Markets

Source: https://www.motu.nz/assets/FacebookImages/How-it-works-2018.jpg

Surprisingly, this lobby has no plans to suspend, mitigate, or reverse climate change via carbon trading or any other means except for monetary gains. The primary aim of the Kyoto Protocol, which was agreed upon during the Convention on Climate Change, was to create a system that would allow for the trade of emission permits. Since then, there have been ongoing attempts to create a positive public perception of the industry. The Paris Agreement, signed on December 12, 2015, by 196 countries, was a legally binding agreement that followed the Kyoto Protocol of 1997.

There are two types of emissions that can be traded on the carbon market (Figure 10): carbon credits and carbon offsets. These terms are sometimes used interchangeably, but they function differently. Carbon credits are permits that allow companies to emit a certain amount of carbon, which are typically bought from the government. Carbon offsets, on the other hand, are obtained by removing a unit of carbon from the atmosphere as part of a company's normal operations. Companies can then sell these offsets to other companies to help reduce their carbon footprint. The emission target, established under cap-and-trade in a number of nations, including Canada, the EU, the UK, China, New Zealand, Japan, and South Korea, determines the issuance of credits. Companies are encouraged to decrease their carbon emissions and remain within their designated limit.

Taking advantage of the high prices of carbon offsets, there have been instances of fraudulent carbon trading, where carbon credits are traded fraudulently. There have been multiple reports of scams related to carbon trading, in which companies sold fake carbon credits or made false assertions about their carbon offsetting activities.

Green energy fraud is the act of making false claims about producing renewable energy, such as solar or wind power, with the aim of receiving government subsidies or attracting investors. There have been numerous reports of green energy fraud in which companies claimed to generate renewable energy while, in reality, using fossil fuels.

E-waste dumping, which involves exporting electronic waste to developing countries where it is often disposed of in harmful and ecologically destructive ways, has been criticised for causing health issues and environmental harm in these countries. While it is often justified in developed countries as a way to decrease waste, this practice has been controversial in developing countries.

6.2 The Lure and Lustre of Greenwashing

Although carbon trading is a viable method for decreasing emissions and encouraging sustainability, it can also be vulnerable to "greenwashing." "Greenwashing" is a marketing strategy that involves making exaggerated or false statements about the ecological advantages of a product or service. This strategy aims to mislead customers into believing that a product is environmentally sustainable or friendly when it is not. Firms may engage in "greenwashing" to profit from the increasing demand for eco-friendly items or to distract attention from their harmful environmental consequences. Greenwashing can take several forms, from using ambiguous or deceptive language in advertisements to highlighting minor environmental features while disregarding more significant ecological impacts.

Products may be labelled with vague or meaningless terms such as "green", "natural", or "eco-friendly" without any certification or evidence to support these claims. Some businesses may use certifications that have nothing to do with the environmental impact of their products, such as "certified organic" labels on non-sustainable products. Companies may also focus on minor environmental benefits while disregarding more severe environmental consequences or highlight one benefit while ignoring other negative environmental impacts. For example, a car may be marketed as "fuel-efficient" while disregarding its high greenhouse gas emissions. Additionally, some companies may use images or language in their advertising to imply environmental benefits that do not exist, such as including images of nature or wildlife to imply that their products are environmentally friendly, even if they are not. This marketing strategy, known as "greenwashing", can be used to take advantage of the growing demand for environmentally friendly products or distract from the negative environmental impact of a company's operations.

In recent times, there has been an increase in instances and examples of "greenwashing," which occurs when businesses attempt to benefit from consumers' growing environmental concerns. Greenwashing can take various forms, ranging from misleading environmental claims to outright lies about a company's sustainability practices. Unfortunately, such practices not only deceive consumers but also have the potential to harm the environment by diverting attention and resources away from genuinely sustainable practices. Consequently, governments and consumer protection agencies worldwide have taken legal action against companies that engage in "greenwashing" in order to hold them accountable and discourage others from following suit. As a result of these legal actions, greenwashing companies have faced fines, penalties, and even court-ordered changes to their environmental practices, demonstrating the severity of the repercussions of greenwashing.

"Greenwashing has numerous examples and case studies.

Volkswagen was discovered to have installed emissions-cheating software in its diesel cars in 2015, allowing them to emit up to 40 times the legal limit of pollutants. Volkswagen misrepresented the environmental friendliness of these vehicles by marketing them as "clean diesel", despite the fact that they were not.

H&M was criticised in 2019 for its "Conscious Collection", which was marketed as eco-friendly. It was discovered, however, that the collection contained non-sustainable materials, such as conventional cotton, and was manufactured in factories with poor working conditions.

In 2020, Nestle was criticised for marketing a new line of KitKat chocolate bars as eco-friendly because they were packaged in recyclable paper. However, it was discovered that the paper used in the preponderance of UK locations where the bars were sold was not recyclable.

2010 saw the introduction of PepsiCo's "Dream Machine" campaign, which encouraged consumers to recycle their beverage containers in special vending machines. It was discovered, however, that the machinery did not recycle the containers but instead dumped them in landfills.

In 2015, ECM Biofilms, a manufacturer of plastic additives, was involved in a settlement with the Federal Trade Commission (FTC). The settlement came about due to false and deceptive claims made by ECM Biofilms regarding the biodegradability of its products when used in plastic items. As part of the settlement, ECM Biofilms was ordered to pay a civil penalty of $70,000 and was prohibited from making similar claims in the future unless it had the appropriate evidence to substantiate them.

The California Attorney General's Office sued JBS USA, a company that processes meat, in 2017 for making false claims about the sustainability of its meat products. The lawsuit alleged that JBS USA's advertising was misleading and did not have sufficient scientific support. The investigation into this matter is still ongoing.

In 2018, the FTC reached a resolution with three companies, namely, American Plastic Manufacturing, Evergreen Plastics, and PDK, for promoting false and deceptive claims about the eco-friendly features of their plastic products. The three companies were collectively fined $180,000 and were prohibited from making comparable assertions in the future without adequate verification.

In 2019, the Federal Trade Commission (FTC) came to an agreement with Chemence, Inc., a producer of adhesive products, for promoting false and unsupported assertions regarding the biodegradability and compostability of its merchandise. As part of the settlement, Chemence was compelled to pay a civil fine of $1.2 million and was prohibited from making similar claims in the future without adequate verification.

There have been many instances of Patanjali, an Indian consumer goods company, potentially engaging in greenwashing. In 2016, the company faced scrutiny when it claimed that its noodles brand, Patanjali Atta Noodles, contained "no added MSG" (monosodium glutamate). Laboratory tests conducted by the Food Safety and Standards Authority of India (FSSAI) later revealed the presence of MSG in the product.

Greenwashing is merely one of the unethical behaviours carried out under the pretext of addressing climate change. Due to the pressing need to tackle climate change, some businesses and individuals resort to misleading, manipulative, or other unethical methods, often masked as "environmentalism." To illustrate, some companies may attempt to deflect criticism by highlighting insignificant or superficial adjustments to their practices while continuing to engage in environmentally damaging activities. Others may leverage climate concerns to take advantage of workers, communities, or natural resources or implement policies that undermine civil liberties or social justice. It is imperative to acknowledge these broader patterns of unethical conduct in the context of climate control, to comprehensively address them, and to ensure that environmentalism is not exploited for exploitative or harmful intentions. Forest offsets are credits bought from forest-related projects to neutralise carbon emissions. However, certain projects of this kind have faced censure for uprooting local communities, infringing on indigenous rights, and contributing to deforestation in other regions.

Businesses may engage in "greenwashing" to take advantage of the growing demand for environmentally sustainable products and practices among consumers. The reasons for "greenwashing" can vary but often stem from a desire to increase sales and improve a company's reputation without making meaningful "greenwashing" changes to their environmental practices. Some companies may also be driven by a need to keep up with changing regulations and customer expectations or by a fear of falling behind in a market that values sustainability. Regardless of the motivation, greenwashing can deceive consumers into thinking they are making eco-friendly choices when in reality, they are supporting companies that do little to address environmental concerns such as climate change. This ultimately harms consumers.

Attempts to measure the extent of greenwashing have been made. In 2010, TerraChoice (now UL Environment) conducted a study that revealed that 95% of the "green" products they investigated had at least one case of greenwashing. A study carried out by The Greenwashing Index in 2020 similarly found that more than 90% of the advertisements they analysed included at least one instance of greenwashing.

CHAPTER 7
The Agenda

Some people suspect that efforts to address climate change may have hidden motivations beyond protecting the environment. This portion of the discussion will explore the potential undisclosed goals of climate change initiatives, such as the interests of different stakeholders, the role of the Intergovernmental Panel on Climate Change (IPCC), the role of corporations, and the scope of wealth redistribution as a factor driving climate action. By considering various perspectives, we can gain a deeper understanding of the complex web of interests and incentives that may be driving the dialogue on climate change and the potential consequences of these covert agendas for future generations and our planet.

7.1 Wealth Redistribution-Elite Enrichment Scheme

With the concentration of wealth increasingly favouring a small group while the gap between the rich and poor widens, some individuals believe there is a global agenda to shift wealth from the less privileged to the wealthy, as exemplified during the COVID-19 pandemic. Consequently, concerns arise about whether climate change policies could exacerbate economic inequalities that benefit the affluent at the expense of the poor. Therefore, it is crucial to closely monitor and address any potential economic disparities that may arise from the implementation of climate change policies.

It is important to investigate how climate change policies, regardless of their causes, effects, or consequences, could lead to the transfer of wealth from disadvantaged individuals to wealthier ones. This investigation entails designing policies and practices that may result in the redistribution of wealth from low-income groups to high-income individuals or corporations through means such as tax policies, government subsidies, or economic policies favouring the affluent. Such an investigation should be approached from a conceptual and theoretical standpoint.

Efforts to tackle climate change will necessitate significant changes in energy production, consumption, and land use management. These changes can have economic ramifications, such as increased costs for businesses that emit greenhouse gases, as well as opportunities for those involved in clean energy production or climate adaptation services. It is vital to remain vigilant for any signs of wealth redistribution throughout the entire process of policy development, monitoring, and implementation, including potential attempts by industries to shift their costs onto the public.

Policies aimed at mitigating climate change may have distributional effects, leading to increased costs for households and businesses reliant on fossil fuels or located in climate-affected regions. However, such policies can also generate new prospects for economic development and job creation, particularly in the fields of clean energy technology and sustainable land management practices.

While evaluating policies and decisions from a theoretical standpoint is important, it is equally crucial to consider their impacts on different demographic groups and their economic well-being. Certain policies may indeed result in higher costs for households or industries heavily dependent on fossil fuels or situated in vulnerable areas. Nevertheless, these policies may also foster opportunities for economic growth and job creation, particularly in the advancement of clean energy technologies and sustainable land management practices.

Claims of a clandestine conspiracy to transfer wealth from the poor to the rich under the guise of addressing climate change are difficult to substantiate due to a lack of credible evidence or reputable sources. Nevertheless, it is important to remain vigilant against powerful corporate and economic entities that may be operating covertly and lacking transparency or accountability.

Challenging the influence of powerful corporate and economic groups clandestinely transferring wealth from the poor to the rich in the name of climate change is a challenging endeavour. It is essential to consider the impact of climate change policies on various demographics and their financial well-being. Simultaneously, evaluating policies and decisions based on their potential impacts and distribution effects is crucial from a conceptual and theoretical perspective.

The challenge of wealth redistribution from the poor to the rich within the context of climate change is complex and lacks a straightforward solution. Policies such as carbon pricing or cap-and-trade systems may lead to a redistribution of wealth from the poor to the wealthy. For example, carbon pricing, which involves imposing a tax on carbon emissions, could result in higher energy costs for consumers, disproportionately affecting low-income households. However, companies that comply with emission regulations and decrease their carbon footprint could reap financial benefits. A possible consequence of this is that financially disadvantaged consumers may end up paying wealthier businesses.

It is possible that a cap-and-trade system, in which businesses are given carbon emission allowances, would have the same effect. Businesses that are able to reduce their emissions below their cap may be able to sell their surplus permits to other businesses that have emitted more than they are allowed.

Certain individuals or organisations may exploit the debate around climate change to advance their own interests or agendas. For instance, the fossil fuel industry may have a financial stake in downplaying the severity of climate change and opposing policies aimed at reducing carbon emissions. Some groups or individuals may also sensationalise the dangers of climate change or propose unrealistic solutions to gain attention and financial backing or achieve their own objectives. In addition, some groups or individuals may attempt to profit from the development of climate change adaptation and mitigation technologies and services. Politicians or political organisations may also use climate change to advance their objectives and sway public opinion.

Climate change may have a significant impact on the insurance industry, which could result in higher insurance rates. The increased frequency of extreme weather events caused by climate change could result in higher payouts by insurance companies for property damage and other losses. The National Bureau of Economic Research published a report in 2020 indicating that the insurance industry paid $18 billion in claims in 2017 following Hurricane Harvey, one of the costliest natural disasters in US history. Additionally, there may be indirect costs associated with climate change, such as reduced economic activity or falling property values, which could also affect the insurance industry. In response, some insurers may use climate change as a justification for increasing their rates.

It seems that an opinion is being formed in favour of climate change. It's extremely challenging to offer a rebuttal in such a setting. Instead of continuing to ask questions, you may find yourself defensively fending off the onslaught of responses and opinions from newcomers with a superficial understanding at best.

Remember that many renewable energy pioneers are also market leaders in the fuel industry. Their ownership structure consists of a complex web of holdings and investors, the unravelling of which is difficult. There is a common misconception that a low-carbon economy will only benefit the renewable energy industry, the energy efficiency industry, and the electric vehicle manufacturing industry. They qualify for special government aid programmes and rebates. To an untrained observer, it may not be immediately apparent, but the continued consolidation of the market and the removal of opportunities for new competitors would lead to a monopolistic situation. This raises legitimate concerns about the potential for monopolisation.

The structure of a company's ownership can have an impact on the company's actions and priorities. Some investors may prioritise short-term profits over the company's long-term viability, while others may give more consideration to environmental, social, and governance factors. However, investors' only motive is profit. Additionally, a small number of people and families frequently own businesses involved in the production and use of fossil fuels, which can have an impact on their political influence and lobbying activities.

In spite of investors' profit-oriented objectives, numerous organisations have emerged with an apparently clear mission to advance sustainable business practices and tackle social and environmental issues. However, in actual practice, these bodies are lobby groups with the sole motive of profiteering.

One organisation that exemplifies this is Business for Social Responsibility (BSR), established in 1992 and headquartered in San Francisco, California. BSR collaborates with companies across diverse industries, including technology, consumer goods, and finance, to identify and implement sustainable business solutions. The board of directors of BSR comprises leaders from the corporate, academic, and nonprofit sectors, reflecting a commitment to a multi-stakeholder approach to sustainability. However, it is crucial to maintain caution regarding the motivations of such organisations because the larger economic and political systems in which they operate continue to have an impact on their actions.

BSR's current board of directors includes:

- Aditi Mohapatra, Chief Sustainability Officer, Wipro Limited
- Andrew Winston, Founder, Winston Eco-Strategies
- Aron Cramer, President and CEO, Business for Social Responsibility

- Carol Cone, CEO, Carol Cone ON PURPOSE
- David Blood, Co-Founder and Senior Partner, Generation Investment Management
- Erin Meezan, Vice President and Chief Sustainability Officer, Interface, Inc.
- Hiro Mizuno, UN Special Envoy on Innovative Finance and Sustainable Investment, Principles for Responsible Investment (PRI)
- Karina Litvack, Non-Executive Director, Eni S.P.A.
- Mark Kramer, Co-Founder and Managing Director, FSG
- Peter Bakker, President and CEO World Business Council for Sustainable Development
- Ravi Saligram, CEO and Director, Newell Brands
- Sarah Middleton, Executive Director, CDP North America
- Syed Ali, President and CEO, Cavium, Inc.

Besides this, there are numerous other such organisations with similar objectives and working on similar lines, which are discussed in the following pages.

The World Economic Forum (WEF), a private economic group, appears to be aiming to take an active role in addressing climate change by bringing together leaders from various sectors, including business, government, civil society, and academia, to discuss and find solutions to complex global challenges. However, it would be unwise to believe what they say because their lifestyle and energy consumption habits show that their main objective is to make money for their partners. Despite this, they have cleverly disguised themselves by labelling themselves as a non-profit organisation that works with companies to promote sustainable business practices and address social and environmental challenges.

The Ceres Coalition is a nonprofit organisation and coalition of investors, companies, and public interest groups working to advance sustainable business practices and address global sustainability challenges, including climate change, water scarcity, and human rights. Ceres was founded in 1989 and is based in Boston, Massachusetts. The current members of Ceres' board of directors include Mindy Lubber, CEO and President of Ceres; Peter Bakker, President and CEO of World Business Council for Sustainable Development; Lisa Jackson, Vice President, Environment, Policy and Social Initiatives, Apple Inc.; Gary Hirshberg, Co-Founder and Chief Organic Optimist, Stonyfield Farm; Veena Ramani, Senior Program Director, Capital Market Systems, Ceres; Susan Baker, Managing Director, Trillium Asset Management; David Blood, Co-Founder and Senior Partner, Generation Investment Management; Daniele Horton, Founder and President, Verdani Partners; Eileen Fisher, Founder and Chairwoman, EILEEN FISHER, Inc.; Scott Mather, Chief Investment Officer, U.S. Core Strategies, PIMCO; Mindy S. Grossman, CEO and Director, WW International, Inc.; Jacqueline Novogratz, Founder and CEO, Acumen and Cheryl Carolus, Board Chair, African Parks Network

The Carbon Pricing Leadership Coalition (CPLC) is a voluntary partnership of governments, businesses, and civil society organisations that work together to promote the use of carbon pricing as a means of reducing greenhouse gas emissions and addressing climate change. The CPLC was launched in 2015 by the World Bank Group, and it has since grown to include more than 35 national and subnational governments, over 160 businesses and investors, and numerous civil society organisations. The CPLC is guided by a steering committee that includes representatives from government, business, and civil society. The current members of the CPLC Steering Committee are the World Bank Group, Canada, Chile, Costa Rica, Ethiopia, France, Germany, Japan, Mexico, New Zealand, the Republic of Korea, Spain, Sweden, the United Kingdom, California, the Chilean Greenhouse Gas Inventory and Reporting Programme, Citi, EDF, Enel, Engie, ERM, ExxonMobil, HSBC, Iberdrola, Mahindra Group, Microsoft, Nestle, Shell, Snam, Suez, Unilever, and the "We Mean Business" group at the World Economic Forum (WEF).

The International Emissions Trading Association (IETA) is a nonprofit business association that promotes market-based solutions to address climate change. IETA was established in 1999 and is headquartered in Geneva, Switzerland. IETA is governed by a board of directors responsible for setting the organisation's strategic direction and overseeing its operations. The current members of IETA's board of directors are Dirk Forrister (President and CEO), Stefano De Clara (Managing Director), Duncan van Bergen (Chairman), Julian Richardson, Marco Alverà, Matthew Gray, Maria Carolina Schmidt, Kazuhisa Koakutsu, Barbara Buchner, Amy Holm, Annette Loske, and Scott Foster (ex officio).

The We Mean Business Coalition is a global nonprofit organisation that brings together businesses, investors, and policymakers to accelerate the transition to a low-carbon economy. The coalition was launched in 2014 and is based in London, UK. The We Mean Business coalition is led by a board of directors responsible for setting the organisation's strategic direction and overseeing its operations. The current members of the board include Paul Simpson, CEO of CDP; Nigel Topping, High-Level Climate Action Champion for COP26; Peter Bakker, President and CEO of the World Business Council for Sustainable Development; Helen Clarkson, CEO of the Climate Group; Sanda Ojiambo, Executive Director and CEO of the United Nations Global Compact; Kirsty Jenkinson, Managing Director, Sustainable Investing at J.P. Morgan Asset Management; and Mafalda Duarte, CEO of the Climate Investment Funds.

The Global Reporting Initiative (GRI) is an international independent standards organisation that helps businesses, governments, and other organisations understand and communicate their environmental, social, and governance (ESG) impacts. The organisation was founded in 1997 by the Coalition for Environmentally Responsible Economies (CERES) and is based in Amsterdam, Netherlands. The GRI develops and maintains the most widely used global standards for sustainability reporting, known as the GRI Standards. These standards provide a framework for organisations to report on a range of ESG issues, including climate change, human rights, labour practices, and supply chain sustainability. The GRI is governed by a multi-stakeholder council, which includes representatives from businesses, civil society organisations, labour unions, and other groups. The council is responsible for overseeing the organisation's strategic direction and ensuring that its activities remain transparent and accountable to stakeholders. The GRI is led by a board of directors responsible for setting the organisation's overall strategy and overseeing its operations. The current members of the GRI board of directors include Eric Hespenheide (Chair), Board Member of GRI, Global Leader of Sustainability Services, Deloitte Touche Tohmatsu Limited; Esther Speck (Vice-Chair), Senior Manager, Group Sustainability, Deutsche Post DHL Group; Paul Druckman (Treasurer), Chair, The Prince's Accounting for Sustainability Project (A4S) Advisory Council; Erika Karp, Founder and CEO, Cornerstone Capital Inc.; Christianna Wood, Non-Executive Director and Chair of the Remuneration Committee, GlaxoSmithKline plc; Michele Lemmens, Chief Investment Officer, MN; Sakie Tachibana, Group Chief Sustainability Officer and General Manager, Corporate Sustainability Division, Ricoh Co., Ltd. and Gerhard Wagner, Chief Financial Officer, Stora Enso Oyj.

The World Business Council for Sustainable Development (WBCSD) is an organisation of over 200 leading companies working together to accelerate the transition to a sustainable world. The organisation was founded in 1995 and is headquartered in Geneva, Switzerland. The WBCSD is governed by a board of directors, which is responsible for setting the organisation's overall strategy and overseeing its operations. The current members of the WBCSD board of directors include Sunny Verghese (Chair), Co-Founder and Group CEO of Olam International Limited; Peter Bakker (Vice-Chair), President and CEO, WBCSD; Jean-Pierre Clamadieu, Chairman of the Board; Engie SA; Ignacio

Galán, Chairman and CEO, Iberdrola SA; Mark Gough, Executive Director, Natural Capital Coalition; Nancy Kete, Senior Advisor, Business and Environment Program, World Resources Institute; Maria Mendiluce, CEO, We Mean Business; Lise Kingo, CEO and Executive Director, United Nations Global Compact; Helena Helmersson, CEO, H&M Group; Jyrki Katainen, President, Sitra; Nizar Al-Adsani, Deputy Chairman and CEO, Kuwait Petroleum Corporation; Martin Brudermüller, Chairman of the Board of Executive Directors, BASF SE; Paul Polman, Co-Founder and Chair, Imagine and Tania Cosentino, President, Schneider Electric Brazil.

The Climate Group and governments to accelerate the transition to a low-carbon economy. The organisation was founded in 2004 and is headquartered in London, UK. The Climate Group is governed by a board of directors, which is responsible for overseeing the organisation's strategic direction and operations. The current members of The Climate Group board of directors include Nick Robins (Chair), Professor in Practice for Sustainable Finance, Grantham Research Institute, London School of Economics; Amy Davidsen (Vice Chair), Executive Director, North America, The Climate Group; Nick Akins, Chairman, President and CEO, American Electric Power; Feike Sijbesma, Honorary Chairman, Royal DSM; Gary Hattem, Managing Director, Head of Global Social Impact, Deutsche Bank; Gil Quiniones, President and CEO, New York Power Authority; Jules Kortenhorst, CEO, Rocky Mountain Institute; Kumi Naidoo, Secretary General, Amnesty International; Lisa Jackson, Vice President, Environment, Policy and Social Initiatives, Apple; Lord Barker of Battle, Executive Chairman, En+ Group; Maria Mendiluce, CEO, We Mean Business and Nigel Topping, CEO, We Mean Business.

Besides these economic pressure groups, several groups are composed entirely of investors and claim to be focused on sustainable investing and addressing climate change. Some examples are:

The Investor Network on Climate Risk (INCR) is a network of institutional investors focused on addressing the financial risks and opportunities of climate change. The network was founded in 2003 and is part of the nonprofit organisation Ceres, which works with investors and companies to promote sustainable business practices. The INCR is governed by a steering committee responsible for setting the organisation's strategic direction and priorities. The current members of the INCR steering committee include Mindy Lubber (Chair), CEO and President, Ceres; Anne Simpson (Vice Chair), Managing Investment Director, Board Governance and Sustainability, California Public Employees' Retirement System (CalPERS); Lauren Compere, Managing Director, Boston Common Asset Management; Lee Wasserman, Director, Rockefeller Family Fund; Mark Fulton, Founding Partner, Energy Transition Advisors; Michael Garland, Assistant Comptroller for Corporate Governance and Responsible Investment, New York City Comptroller's Office; Scott Mather, CIO US Core Strategies, PPIMCO; and Veena Ramani, Senior Programme Director, Ceres.

The Principles for Responsible Investment (PRI) is an international network of investors focused on promoting responsible investment practices. The organisation was founded in 2006 by a group of investors collaborating with the United Nations. The PRI is governed by a board of directors, which is responsible for overseeing the organisation's strategy and operations. The current members of the PRI board of directors include Martin Skancke (Chair), Chair of the Board, Storebrand Asset Management; Masja Zandbergen-Albers (Vice Chair), Head of Sustainability Integration, Robeco; Sandra Carlisle, Head of Responsible Investment, HSBC Global Asset Management; Catherine Howarth, CEO, ShareAction; Hiro Mizuno, Managing Director and CIO, GPIF; Kristian Fok, CIO, Cbus Super Fund; Mark Machin, President and CEO, Canada Pension Plan Investment Board; Peter Sandahl, CEO, P+ Investments; Ron Mock, President and CEO, Ontario Teachers' Pension Plan; and Tanya Carmichael, Senior Advisor, Responsible Investment, AustralianSuper.

The Carbon Tracker Initiative is a non-profit organisation focused on aligning financial markets with the transition to a low-carbon economy. The organisation was founded in 2011 by a group of financial analysts and environmentalists. The Carbon Tracker Initiative is governed by a board of directors, which is responsible for setting the organisation's strategy and overseeing its operations. The current members of the Carbon Tracker Initiative board of directors include Jeremy Leggett (Chair), Founder and Director, Solarcentury and Chairman, Carbon Tracker Initiative; Faith Ward (Vice Chair), Chief Responsible Investment Officer, Brunel Pension Partnership; Henrik Jeppesen, Head of Investor Outreach, Carbon Tracker Initiative; James Cameron, Founder and Chairman, Overseas Development Institute and Senior Adviser, SYSTEMIQ; Mark Campanale, Founder and Executive Director, Carbon Tracker Initiative; Nick Robins, Professor in Practice, Sustainable Finance, London School of Economics and Political Science; Sarah Breeden, Executive Director, International Banks Supervision, Bank of England and Tom Burke, Chairman, E3G.

The Climate Bonds Initiative is an international non-profit organisation focused on mobilising the bond market for climate change solutions. The organisation was founded in 2011 and is headquartered in London, UK. The Climate Bonds Initiative is governed by a board of directors, which is responsible for overseeing the organisation's strategy and operations. The current members of the Climate Bonds Initiative board of directors include Sean Kidney (CEO and Co-founder), Climate Bonds Initiative; Justine Leigh-Bell (Chair), Independent Consultant; Helena Viñes Fiestas (Vice-Chair), Global Head of Stewardship and Policy, BNP Paribas Asset Management; Axel van Nederveen, Global Head of Fixed Income Research, Amundi; Ben Caldecott, Founding Director, Oxford Sustainable Finance Programme, University of Oxford; Brad Church, Managing Director, Global Head of Green Bonds, Citi; Cecilia Repinski, Director of Environmental, Social and Governance (ESG) Research, MSCI; Christiana Figueres, Co-founder, Global

Optimism and Former Executive Secretary, UN Framework Convention on Climate Change; Fabrizio Zago, Deputy Head of Group Public Affairs and Sustainable Impact, AXA; Lila Karbassi, Chief of Programmes, UN Global Compact; Mamadou-Abou Sarr, CEO and Founder, V-Square Quantitative Management; Samantha Harris, Director, Climate Investment Funds and Sean Kidney, CEO and Co-founder, Climate Bonds Initiative.

The Institutional Investors Group on Climate Change (IIGCC) is a European membership organisation of investors that is focused on addressing the risks and opportunities associated with climate change. The organisation was founded in 2001 and is based in London, UK. The IIGCC is governed by a board of directors, which is responsible for setting the organisation's strategy and overseeing its operations. The current members of the IIGCC board of directors include Peter Damgaard Jensen (Chair), CEO, PKA A/S; Philippe Desfossés (Vice Chair), CEO, ERAFP; Stephanie Maier (Vice Chair), Director of Responsible Investment, HSBC Global Asset Management; Frederic Samama (Treasurer), Deputy Global Head of Institutional & Sovereign Clients, Amundi Asset Management; Mark Lewis, Global Head of Sustainability Research, BNP Paribas Asset Management; John McKinley, Head of Institutional Relationship Management Europe, BlackRock; Caroline Stolz, Head of Investment Strategy & Risk Management, Zurich Insurance Group; Thomas Tayler, Senior Portfolio Manager, Sustainable & Responsible Investments, Legal & General Investment Management and Kristian Fok, CEO, Storebrand Asset Management.

The Interfaith Centre on Corporate Responsibility (ICCR) is a non-profit organisation based in New York City, USA. It was founded in 1971 and is made up of over 300 member organisations, including faith-based institutional investors, foundations, asset management companies, and other socially responsible investors. The ICRC's mission is to mobilise the collective influence of its members to promote corporate responsibility and sustainable practices. The organisation engages with companies through shareholder advocacy, dialogue, and other forms of engagement to promote issues related to environmental sustainability, social justice, and corporate governance. The ICCR is governed by a board of directors, which is composed of representatives from member organisations. The board is responsible for setting the organisation's strategic direction and overseeing its operations. The current members of the ICCR board of directors include Sister Nora Nash (Chair), Sisters of St. Francis of Philadelphia; Fr. Séamus Finn (Vice-Chair), OMI, Missionary Oblates of Mary Immaculate; Kathy Mulvey (Treasurer), ESG Strategy and Communications Consultant; Julie Tanner (Secretary), Christian Brothers Investment Services; David Schilling, ICCR Senior Programme Director; Sr. Barbara Aires, Sisters of Charity of Saint Elizabeth; Tom McCaney, Mercy Investment Services; Michael Passoff, Proxy Impact; and Fr. Michael Crosby, OFM Cap, Province of St. Joseph of the Capuchin Order.

The Shareholder Association for Research and Education (SHARE) is a Canadian non-profit organisation that works to promote responsible investment practices by institutional investors. SHARE was founded in 2000 and is headquartered in Vancouver, British Columbia, with offices in Toronto and Montreal. SHARE's mission is to support institutional investors in integrating environmental, social, and governance (ESG) factors into their investment decision-making and ownership practices. The organisation works with pension funds, foundations, endowments, and other institutional investors to promote sustainable investment practices through shareholder engagement, proxy voting, and other forms of advocacy. SHARE is governed by a board of directors, which is composed of representatives from member organisations. The board is responsible for setting the organisation's strategic direction and overseeing its operations. The current members of the SHARE board of directors include Dale Richmond (Chair), British Columbia Teachers' Federation;

Robb Zuk (Vice-Chair), AMUN AG; Irene Nattel (Secretary-Treasurer), RBC Dominion Securities Inc.; Ian Robertson, University of Toronto Asset Management; Jean-Philippe Renaut, Bâtirente; Marlene Puffer, Alignvest Investment Management; Hélène Poirier, Fonds de solidarité FTQ; and Heather Keachie, New Brunswick Investment Management Corporation.

The Investor Group on Climate Change (IGCC) is a collaboration of investors across Australia, and New Zealand focused on addressing the risks and opportunities associated with climate change. The group was established in 2003 and now has over 80 members, including institutional investors, fund managers, and others with investments in the region. The IGCC's mission is to encourage and support its members to integrate climate change into their investment decision-making and engage with companies, policymakers, and other stakeholders on the issue. The group works to raise awareness of the financial risks associated with climate change and the opportunities for investing in a low-carbon economy. The IGCC is governed by a board of directors, which is composed of representatives from member organisations. The board is responsible for setting the group's strategic direction and overseeing its operations. The current members of the IGCC board of directors include Emma Herd (Chair), IGCC; Adam Kirkman (Deputy Chair), AustralianSuper; Richard Brandweiner, Pendal Group; Andrew Gray, Australian Ethical Investment; Vicky Hyde-Smith, New Zealand Super Fund; Will MacAulay, Cbus; David Macri, HESTA; Alison Martin, Zurich Australia; Kirsten Moller, QBE; and Nathan Fabian (ex officio), IGCC.

CDP (formerly known as the Carbon Disclosure Project) is a global non-profit organisation that runs a global disclosure system for companies, cities, states, and regions to manage their environmental impacts. The organisation was founded in 2000 and is headquartered in London, UK, with offices in over 50 countries. CDP's main aim is to encourage companies, cities, states, and regions to disclose their environmental impact data, including information on their carbon emissions, water usage, and forest management practices. The organisation believes that making this information publicly available can help drive action on climate change and other environmental issues. CDP is governed by a board of trustees, which is composed of representatives from its member organisations. The board is responsible for setting up CDP's strategy and overseeing its operations. The current members of the CDP board of trustees include Tariq Fancy (Chair), The Rumie Initiative; Paul Dickinson (Founder and Executive Chair), CDP; Frances Way (Chief Strategy Officer), CDP; Lisa Walker (CEO), Ecosphere+; Jennifer Austin, Credit Suisse; Marilyn Ceci, JP Morgan Chase; Tim Mohin, Persefoni; David Pitt-Watson, London Business School; and Paula DiPerna, CDP North America.

These organisations represent a growing movement of investors who claim to recognise the risks and opportunities associated with climate change on the pretext of integrating environmental, social, and governance considerations into their investment decisions. By engaging with companies and policymakers, these groups seek to drive sustainable business practices and accelerate the transition to a low-carbon economy.

However, it is important to remember that these organisations are typically governed by a board of directors or trustees, and decisions are made through a collaborative process involving multiple stakeholders rather than merely the individuals involved. Individuals' identities are concealed through a network of trusts, philanthropic organisations, and for-profit businesses. It is extremely difficult to extract information about the individual owners of these companies, despite the fact that this information is ostensibly available on their websites, annual reports, and other disclosure statements. These individuals do not use LinkedIn and other professional networking sites to hide their identities.

Upon examining the members and affiliations of these economic groups, it becomes evident that both the individuals and the organisations they represent have deep corporate ties and interconnections. Many of these individuals hold positions on multiple boards, resulting in a complex web of overlapping groups that are difficult to distinguish from one another. The fundamental issue at hand is that these members and organisations act on behalf of major corporations, leaving no doubt that they are lobbying groups whose main objective is to safeguard the interests of their employers and influence government policies to benefit their commercial operations.

The corporate ties and interconnections of the members and organisations in economic groups are significant because they highlight a potential motive for their lobbying efforts, which is to benefit their masters and commercial operations. The issues of wealth redistribution often exacerbate the problem by enriching the already wealthy while corporate avarice continues to prevail.

7.2 Corporate Avarice

The standard legal procedure for civil, financial, or criminal investigations typically involves considering individuals or entities with a potential benefit or vested interest in the crime or proceedings. These parties, closely connected to the crime, are often regarded as having a "motive" for committing it. For instance, in cases of financial fraud, investigators may scrutinise individuals who could gain financially from the fraudulent activity. In civil cases, they may examine those with a direct stake in the outcome, while criminal investigations focus on individuals who may have had a motive to commit the crime.

While it is true that suspicion often arises from examining motives during investigations, it is crucial to acknowledge that motive alone is insufficient to establish guilt. To prove someone's involvement in a crime or their benefit from it, investigators must gather and analyse various forms of evidence, including physical evidence, financial records, witness testimonies, and other relevant information related to the case. For example, in a financial fraud case, investigators may delve into bank statements, emails, and other financial documents to trace the money flow and identify potential beneficiaries of the fraud. In a criminal investigation, they may collect DNA evidence or fingerprints from the crime scene to identify possible suspects.

Investigating motives can aid in uncovering the underlying motivations or agendas behind someone's actions. By identifying what someone stands to gain from a specific action, investigators can better comprehend their intentions and motivations, which can be valuable in determining whether a crime has been committed and identifying potential culprits.

Applying these principles, let us now investigate who stands to benefit from the climate change hype. One group that may benefit from the hype around climate change is the renewable energy industry, which includes solar, wind, and hydroelectric power companies. There are many private companies involved in renewable energy, including solar, wind, and hydroelectric power. But major players are Brookfield Renewable, Enphase Energy, Orsted, Siemens Gamesa, SunPower, Tesla, and Vestas. Now let us look at these companies individually and their ownership structures one by one.

Brookfield Renewable is a Canadian company that develops, owns, and operates hydroelectric, wind, and solar power facilities in North and South America, Europe, and Asia. Brookfield Renewables is a publicly traded company listed on the New York and Toronto Stock Exchanges. Brookfield Asset Management, a global alternative asset management company, is the majority owner of Brookfield Renewable and, as of 2021, owns approximately 50% of the outstanding shares of Brookfield Renewable. Other significant shareholders included institutional investors like Vanguard, BlackRock, and State Street.

Enphase Energy designs and manufactures microinverters, which are used in solar power systems to convert DC power to AC power for use in homes and businesses. Enphase Energy is a publicly traded company listed on the Nasdaq Stock Exchange. As of 2021, the largest shareholder of Enphase Energy was Vanguard Group, Inc., an American investment management company. Vanguard owned approximately 11.8% of the outstanding shares of Enphase Energy. Other significant shareholders included BlackRock, Inc., Wellington Management Group, and The Capital Group Companies, Inc.

Orsted is a Danish company that specialises in offshore wind power. They develop, build, and operate wind farms in Europe, North America, and Asia. Orsted is a publicly traded company listed on the Nasdaq Copenhagen Stock Exchange. As of 2021, the largest shareholder of Orsted was the Danish government, which owned approximately 50.1% of the outstanding shares. Other significant shareholders included institutional investors like BlackRock, Norges Bank Investment Management, and The Vanguard Group, Inc.

Siemens Gamesa is a Spanish-German and publicly traded company that designs, manufactures, and instals wind turbines for both onshore and offshore projects. The largest shareholder of Siemens Gamesa is currently Siemens AG, which owns approximately 67% of the company's shares. The remaining shares are publicly traded on stock exchanges such as the Madrid Stock Exchange and the Frankfurt Stock Exchange.

SunPower designs manufactures, and instals solar panels for residential, commercial, and utility-scale projects. They also offer energy storage solutions. Maxeon Solar Technologies, Ltd. is currently the owner of SunPower. In 2019, SunPower spun off from its parent company, Total S.A., and became an independent, publicly traded company under the name Maxeon Solar Technologies. As of 2021, SunPower's largest shareholder was Total Energies, a French multinational energy company, which owned approximately 52% of the company's outstanding shares. Other significant shareholders included BlackRock, Vanguard, and Fidelity Investments, the best-known investment management companies in the world.

Founded in 1975, Vanguard has grown to manage over $ seven trillion in assets as of 2021. Vanguard is known for its low-cost index funds and ETFs and has become a popular choice for individual investors and institutions alike. BlackRock is one of the world's largest asset management companies, with over $ nine trillion in assets under management as of 2021. Founded in 1988, the company has grown to become a major player in the investment management industry, offering a wide range of products and services to individual and institutional investors.

BlackRock and Vanguard are investment management companies with very complex and opaque structures. It is almost impossible to find out who the major individual operatives of these companies are. This complex structure further strengthens the needle of suspicion towards them. As these companies are there just to make money for the sake of making money, they have no other purpose than this. The agenda is straight and simple: the transfer of wealth.

Both Vanguard and BlackRock have equal investments in both sectors, the fossil fuel industry and the renewable energy industry. Any indication of their concern for climate change, if any, falls flat in the light of these facts.

Vanguard is one of the largest asset managers in the world and holds significant investments in various sectors, including the fossil fuel industry. According to its 2020 Annual Report, Vanguard held $78 billion in energy-related investments, including investments in oil and gas companies as well as companies involved in the production and distribution of coal, oil, and gas. However, Vanguard also offers a range of fossil-free investment options for investors who wish to exclude these companies from their portfolios.

Similarly, BlackRock, another major asset manager, has investments in the fossil fuel industry. According to a report from the Institute for Energy Economics and Financial Analysis, BlackRock held investments worth $85 billion in coal, oil, and gas companies as of December 2020. However, like Vanguard,

BlackRock also offers sustainable and ESG-focused investment options for investors who want to invest in companies with a focus on environmental, social, and governance (ESG) factors.

As such, the companies are non-personal entities and cannot act at their own will. It is the individuals behind them who are pulling the strings. It is those individuals who are driving the agenda.

The debate over climate change and the role of fossil fuel companies in exacerbating the problem has been ongoing for decades. Many believe fossil fuel companies are solely interested in maintaining their profits and are not invested in promoting a cleaner, more sustainable future. However, recent research has shown that this argument does not hold water, as the owners and operators of fossil fuels and renewable energy are often the same.

It is true that many of the largest fossil fuel companies have been accused of funding climate change deniers and sceptics, a move that has been seen as a ploy to divert attention from the issue and camouflage their own role in the problem. In fact, many fossil fuel companies are investing heavily in renewable energy technologies and diversifying their portfolios to include more sustainable options. For example, Shell has invested billions of dollars in wind and solar energy projects, while BP has pledged to become carbon neutral by 2050 and is investing in electric vehicle charging infrastructure.

In addition, many of the largest renewable energy companies are owned or operated by the same individuals or entities as their counterparts in the fossil fuel industry. For example, General Electric, one of the largest wind turbine manufacturers in the world, also produces gas turbines and other fossil fuel-related products. Many of the largest utility companies, such as Duke Energy and NextEra Energy, generate power from both renewable and fossil fuel sources.

This shows that the distinction between fossil fuel and renewable energy companies is not as clear-cut as some may believe. The owners and operators of these companies are often the same, and many are actively investing in a cleaner, more sustainable future.

7.3 Green Deals- Wheel Dealing

The Green Deal has been touted as a policy that will help combat climate change by promoting sustainable development and reducing greenhouse gases while leaving low-income households behind.

Moreover, the Green Deal is often criticised for being overly bureaucratic and complex. The policy consists of a vast array of regulations, incentives, and subsidies, which can be confusing and difficult to navigate. This complexity can discourage businesses and consumers from participating in the scheme, which ultimately undermines its effectiveness.

Another issue with the Green Deal is that it places a disproportionate burden on small businesses. The policy requires businesses to reduce their carbon footprint, but many smaller companies simply do not have the resources to make the necessary changes. This puts them at a competitive disadvantage compared to larger corporations, which may be able to afford to implement expensive sustainability measures.

Additionally, there are concerns that the Green Deal is not ambitious enough to address the scale of the climate crisis. While the policy sets out clear targets for reducing emissions, these targets are not sufficiently ambitious to address the enormity of the problem. This has led to accusations that the Green Deal is a "greenwashing" exercise designed to give the appearance of progress while doing little to actually address the underlying issues.

While the Green Deal may have good intentions, it is clear that the policy is not without its flaws. From placing a heavy financial burden on consumers to being overly complex and not ambitious enough, the Green Deal is a trap for wealth transfer that may ultimately do more harm than good. As such, policymakers must take a more nuanced approach when it comes to addressing climate change, one that does not rely solely on policies like the Green Deal.

Real-life case histories also illustrate the challenges facing the Green Deal. For example, a UK government scheme aimed at promoting energy-efficient upgrades to homes under the Green

Deal was scrapped in 2015 after it failed to attract enough participants. Similarly, a study of the Green Deal in Germany found that the policy was not achieving its intended goals due to low levels of participation.

Several countries around the world have implemented policies or initiatives related to a "Green Deal" or similar concept. Here are some examples:

United States: In March 2021, the American Rescue Plan Act included a provision for a "Civilian Climate Corps" as part of a broader plan to address climate change and transition to a clean energy economy. This has been referred to as a "Green New Deal" by some advocates.

European Union: The European Green Deal was announced in December 2019 with the goal of achieving climate neutrality by 2050. The plan includes a range of policies and initiatives related to energy, transportation, agriculture, and other sectors.

United Kingdom: The Green Deal was a policy initiative in the UK from 2013 to 2015 that aimed to encourage homeowners and businesses to invest in energy efficiency measures. The programme was discontinued due to low uptake and criticism of its design.

Canada: In September 2020, the Canadian government announced its intention to introduce a "green recovery" plan to help rebuild the economy in the wake of the COVID-19 pandemic. The plan includes measures related to clean energy, energy efficiency, and other areas.

South Korea: In July 2020, the South Korean government announced a "Green New Deal" that includes a range of policies related to renewable energy, electric vehicles, and other areas. The plan aims to create jobs and reduce greenhouse gas emissions.

France: In January 2020, French President Emmanuel Macron announced a "Green New Deal" for France, with a focus on reducing carbon emissions and promoting sustainable development.

Spain: In May 2020, the Spanish government announced a "Green Deal" as part of its COVID-19 recovery plan. The plan includes measures related to renewable energy, energy efficiency, and sustainable transport.

Germany: The German government has announced plans for a "Green Deal" that would involve investing in clean energy, electric vehicles, and other areas. However, the details of the plan are still being developed.

Australia: In 2019, the Australian Greens political party proposed a "Green New Deal" for Australia, with a focus on renewable energy and sustainable agriculture.

Japan: The Japanese government has announced plans to achieve carbon neutrality by 2050 and is developing policies related to renewable energy, energy efficiency, and other areas. However, it has not used the term "Green Deal" to describe these efforts.

Several countries around the world have implemented policies or initiatives related to a "Green Deal" or similar concept. These include the US, European Union, UK, Canada, South Korea, France, Spain, Germany, Australia, and Japan. The US has implemented a "Civilian Climate Corps" as part of a plan to address climate change and transition to a clean energy economy. The European Union has announced a "Green Deal" with the goal of achieving climate neutrality by 2050. Canada has introduced a "green recovery" plan to help rebuild the economy in the wake of the COVID-19 pandemic.

7.4 IPCC-The Climate Cabal

To understand why the Intergovernmental Panel on Climate Change (IPCC) has taken on the role of attributing human activity to climate change and promoting its views to the public, it is necessary to investigate and analyse the organisation. This includes examining the factors influencing its agenda and the actors driving it. It is also important to understand how scientists are selected and recruited for the working groups, task forces, and expert panels within the IPCC. Furthermore, it is worth investigating the challenges encountered by individuals seeking membership in the IPCC as private individuals, as well as the various factors that influence scientists' choices to either join or abstain from participating.

The Intergovernmental Panel on Climate Change (IPCC) is composed of three main constituent groups: governments, scientists, and observers. The IPCC operates through a structure of working groups, task forces, and expert panels, and its reports are based on the work of three working groups: the Physical Science Basis of Climate Change, Impacts, Adaptation, and Vulnerability, and Mitigation of Climate Change. The selection process for scientists who contribute to the Intergovernmental Panel on Climate Change (IPCC) reports is rigorous and transparent. It includes a nomination process, an eligibility check, an expert review, and approval by the IPCC Plenary. The current members of the IPCC Bureau are Hoesung Lee (Republic of Korea), Thelma Krug (Brazil), Ko Barrett (USA), Youba Sokona (Mali), Panmao Zhai (China), Hans-Otto Pörtner (Germany), Jim Skea (UK), Kiyoto Tanabe (Japan), Eduardo Calvo Buendia (Peru), Lisa Alexander (Australia), Nathalie Olioso (France), Joy Pereira (Malaysia), Debra Roberts (South Africa), Ramón Pichs-Madruga (Cuba), and ex-officio members.

Accessing accurate information from the IPCC can be a daunting task, as it is a large, publicly funded organisation with numerous subsets dedicated to policy, management, and programme implementation. However, there is one subset, the Working Group, that focuses solely on the scientific assessment of climate change. Specifically, IPCC Working Group I (WG I) is responsible for evaluating and reporting on the physical science basis of climate change. This includes analysing the Earth's atmosphere, oceans, and land surface and examining how human activities affect the climate system. Their main objective is to understand the causes, mechanisms, and projected future of climate change, as well as the potential risks and impacts associated with it.

The main challenge and significant constraint with the IPCC are that all of its activities and focus areas are structured, directed, and instructed with the underlying assumption that the science of climate change is fully resolved. This assumption is founded on the consensus among scientists that human activities are the dominant cause of global warming, and it is not open to debate or questioning.

7.5 Protagonist Zealotry

Climate protagonists are individuals or groups that advocate for action to address climate change and work to raise awareness about its impacts on the environment, economy, and society. These protagonists may come from any profession, including scientists, activists, politicians, business leaders, and individuals. It is possible that individuals who exhibit strong zealotry have been swayed, influenced, or manipulated through a process commonly referred to as brainwashing through the use of psychological techniques such as repetitive messaging, emotional manipulation, and social isolation to break down an individual's existing beliefs and replace them with a new set of beliefs or ideologies.

If a person thinks that human activity is to blame for climate change, they can express their opinions in various ways, such as by writing essays, books, or research papers; contacting policymakers; or spreading awareness of the problem. However, it is unlikely that they would spend their entire lives advocating for this cause unless there was some financial gain involved. Those who fall into this category of advocating for financial gain rather than genuine concern may be using climate change as an agenda to earn their livelihood.

I have a strong suspicion that the authors of the paper titled "Quantifying the consensus on anthropogenic global warming in the scientific literature", published in the journal Environmental Research Letters in 2013, and the advocates of that group are the driving force behind the AGW movement. This paper is the foundation of the AGW theory and has played a significant role in shaping public and political discussions on climate change, specifically the consensus among climate scientists that human activities are causing global warming. The results of this study have been cited in numerous subsequent research projects and reports and have influenced public understanding and policy decisions on climate change. It appears that Al Gore is at the forefront of this movement.

The opposite perspective can also be applied to individuals or groups that are sceptical about climate change. These climate sceptics, also referred to as climate change deniers, do not agree with the scientific consensus on climate change and question the evidence of human-induced global warming. They frequently argue that climate change is a natural occurrence that has occurred throughout Earth's history and that the current warming trend may not necessarily be the result of human actions.

The idea that climate sceptics are supported financially by the fossil fuel industry is a common belief, but it is not accurate. The lack of a clear distinction between those who support renewable energy and those who support fossil fuels, as discussed in this book, undermines this argument. Even though they may go by different names, the same people with similar interests ultimately control businesses and companies. The workings of these entities are intricate and complex, making it difficult to untangle their web of deceit that spans multiple corporate layers.

CHAPTER 8

The Future

"Climate Leviathan: A Political Theory of Our Planetary Future" by Mann and Wainwright (Mann, Geoff. Climate Leviathan, 2018) offers a theoretical exploration of the political dynamics surrounding climate change and its implications for global governance. The authors argue that these governance models are shaped by the underlying political and economic structures and interests. They analyse the power dynamics between states, corporations, and civil society in the context of climate change and explore the challenges of achieving collective action on a global scale. Additionally, the book delves into the social and ecological implications of different governance approaches, highlighting the need for democratic and inclusive decision-making processes that address the intersecting challenges of climate change, inequality, and sustainability.

The author presents four scenarios based on the interaction between two key dimensions: the concentration of political power (centralised or decentralised) and the concentration of economic power (concentrated or dispersed). The four scenarios proposed in the book are:

1. Climate Leviathan: In the scenario known as a centralised form of governance with concentrated economic power is envisioned.

It portrays a global authoritarian regime that prioritises stability and control over individual freedoms, emphasising top-down decision-making in tackling climate change. This scenario involves the rise of a potent and centralised global authority or a coalition of states which wields substantial control over policy-making and enforcement actions.

2. Climate Behemoth: In this situation, the distribution of governance is decentralised, but economic power remains centralised. It entails an approach driven by the market, where corporations hold significant influence, relying on market forces and voluntary actions to address climate change. While individual liberties are important, there may be a lack of global coordination. This scenario depicts a type of authoritarian governance not exclusively centred on tackling climate change. The Climate Behemoth symbolises a formidable, undemocratic entity that seizes the opportunity presented by climate change to consolidate and enhance its control over global matters.

3. Climate Mao: This scenario merges decentralised political power with dispersed economic power. It imagines a grassroots movement led by civil society and local communities, operating from the bottom up. The focal points are participatory decision-making, fairness, and environmental justice as fundamental principles for addressing climate change. This scenario envisions an approach that is both decentralised and participatory, with local communities and social movements taking the lead in driving climate action. The emphasis lies on local independence, collective decision-making, and mobilisation from the grassroots level.

4. Climate X: This scenario portrays a world in which political and economic power is widely distributed. It envisions a future where a global democratic governance system arises, highlighting decentralised decision-making, fair allocation of resources, and active involvement from various stakeholders. This scenario delves into a future where global structures of

governance break down, resulting in a fragmented and disorderly world. Without a unified global authority, regional or local actors pursue their individual interests and approaches to addressing climate change.

8.1 The Future of Energy

As the world's population grows and the demand for energy increases, it is crucial to find sustainable and renewable energy sources that can meet our energy needs without damaging the environment. Technology advancements, policy changes, and a growing awareness of the need for sustainable energy sources will all influence the future of energy.

One of the most significant trends in the future of energy is the shift towards renewable energy sources such as solar, wind, hydroelectric, and geothermal. As the cost of renewable energy continues to decrease, more countries are investing in renewable energy infrastructure, and the use of renewable energy is expected to grow.

Another trend in the future of energy is the development of energy storage technologies. Energy storage technologies, such as batteries and fuel cells, will be essential in managing the intermittent nature of renewable energy sources such as solar and wind power. With energy storage technologies, excess energy produced during peak periods can be stored and used when demand is high, ensuring a constant supply of energy.

Advances in smart grid technology will also influence the future of energy. Smart grids use digital technology to manage the distribution of energy, making it more efficient and reducing the risk of power outages. With smart grids, consumers can also monitor their energy usage and adjust their consumption to reduce their energy costs and carbon footprint.

The total energy requirement of the world, also known as global primary energy demand (GPED), is the energy needed to meet the world's energy needs for various purposes, such as electricity generation, transportation, and heating. This energy demand is typically measured in terms of the amount of energy consumed in a given year, and it includes both renewable and non-renewable sources of energy. According to the International Energy Agency's (IEA) World Energy Outlook 2021 report, global primary energy demand was 166,120 TeraWatt-hours (TWh) in 2019. The IEA predicts that global primary energy demand will continue to grow in the coming years, driven by population growth, rising incomes, and increasing urbanisation, particularly in developing countries.

Fossil fuels are currently the dominant source of energy worldwide, accounting for around 84% of global primary energy demand in 2020, according to the BP Statistical Review of World Energy. The estimates based on current production and consumption rates suggest that the world's proven oil, natural gas, and coal reserves may last several decades to a few hundred years. According to the BP Statistical Review of World Energy 2021, the world's proven oil reserves are estimated to last for approximately 48.9 years, proven natural gas reserves for 54.8 years, and proven coal reserves for 133 years.

8.1.1 Renewable Energy

One of the most abundant and widely used forms of renewable energy is solar energy. It is generated from the sun's radiation, which can be captured and converted into electricity using solar panels. Wind energy is another widely used resource where the wind is used through the use of wind turbines.

The force of moving water is what produces hydroelectric power, typically through the use of dams and turbines.

Geothermal energy is generated from the earth's heat, typically by tapping into geothermal reservoirs and using steam or hot water to drive turbines. The movement of ocean currents and tides is what produces tidal energy. Biomass energy is energy generated from organic matter, such as plant material or waste, converted into fuel.

There are many companies and countries involved in renewable energy. Here are some examples of companies and countries that are particularly active in the field of renewable energy.

China is the world's largest producer of solar panels and wind turbines and is investing heavily in other renewable energy technologies such as hydropower, geothermal, and bioenergy. Companies like China National Nuclear Corporation, State Power Investment Corporation, and China Energy Engineering Corporation are among the top renewable energy companies in the country.

The United States is a leader in wind and solar power, and many American companies like NextEra Energy, Tesla, and SunPower are developing innovative renewable energy solutions. The U.S. government also provides tax credits and other incentives for renewable energy development and has set goals to achieve 100% clean energy by 2035.

Germany is a pioneer in renewable energy, with a goal to achieve 80% renewable energy by 2050. Many German companies, such as Siemens, EON, and Innogy, are developing and implementing renewable energy solutions.

India is rapidly expanding its renewable energy capacity, particularly in solar power. The Indian government has set a goal of achieving 450 GW of renewable energy capacity by 2030, and companies like Tata Power, Adani Green Energy, and ReNew Power are among the top renewable energy companies in the country.

Denmark is a leader in wind energy and has set a goal to achieve 100% renewable energy by 2050. Companies like Vestas and Orsted are among the top renewable energy companies in the country.

There are many companies around the world that are involved in the renewable energy industry, including Vestas Wind Systems (Denmark), the world's largest manufacturer of wind turbines, involved in the development and implementation of wind energy projects globally; Enel Green Power (Italy) is a subsidiary of Enel Group and is one of the world's largest renewable energy companies that develops and operates renewable energy projects in a variety of sectors, including wind, solar, hydro, and geothermal; NextEra Energy (USA) is a renewable energy company based in the United States that develops and operates wind, solar, and energy storage projects. It is one of the largest renewable energy companies in the world and has a goal of reaching 30 GW of renewable energy capacity by 2030. Canadian Solar (Canada) is a solar panel manufacturer and developer based in Canada and is one of the largest solar companies in the world. The company develops and operates solar projects globally and provides solar panels and other equipment to other renewable energy companies. Tesla (USA) is primarily known for its electric cars, but the company is also involved in the development and implementation of renewable energy projects. Tesla produces solar panels and energy storage systems and is involved in the development of solar projects in a number of countries. Siemens Gamesa Renewable Energy (Spain) is a global leader in wind turbine manufacturing and has a presence in over 90 countries. The company is involved in the development and implementation of wind energy projects around the world.

Some companies operating in the renewable energy industry may also have operations in the fossil fuel industry, to varying degrees. For example, BP is a global oil and gas company, but in recent years it has started to invest more heavily in renewable energy. The company has set a goal to become a net-zero company by 2050 and has pledged to increase its renewable energy capacity to 50 GW by 2030. Shell is another global oil and gas company that has been investing in renewable energy in recent years. The company has set a goal to achieve net zero emissions by 2050 and has invested in wind and solar projects, as well as electric vehicle charging infrastructure. TotalEnergies is a French energy company that has historically been involved in the fossil fuel industry but has recently increased its investment in renewable energy. The company has set a goal to achieve net zero emissions by 2050 and has a growing portfolio of renewable energy projects, including solar and wind. Eni: Eni is an Italian oil and gas company that is also involved in the development of renewable energy projects. The company has set a goal to achieve net zero emissions by 2050 and has invested in a range of renewable energy technologies, including wind, solar, and bioenergy.

There are a variety of subsidies and incentives available for renewable energy that vary by country, and even within countries, there may be different policies and incentives at the federal, state, provincial, and local levels. Some examples of renewable energy subsidies and incentives in different countries are the United States federal government offers a tax credit of up to 26% of the cost of a renewable energy system, such as solar or wind power, for residential and commercial installations. Some states also offer additional incentives, such as rebates or performance-based incentives; Germany has a feed-in tariff programme that guarantees a fixed price for electricity generated from renewable sources for a certain period. This provides certainty for investors and helps promote the growth of the renewable energy industry. China offers a range of incentives for renewable energy, including tax exemptions and reductions, subsidies for equipment purchases, and preferential access to financing; India has a range of incentives for renewable energy, including a national solar mission that aims to increase the country's solar capacity to 100 GW by 2022 and a range of state-level incentives such as feed-in tariffs, subsidies, and tax exemptions. Japan has a feed-in tariff programme similar to Germany's that provides a guaranteed price for renewable energy generation. The government also offers low-interest loans and tax credits for renewable energy investments.

The specific companies that benefit from renewable energy subsidies and incentives can vary widely depending on the country and the specific policies in place. However, some of the companies that have been major players in the renewable energy industry and have received subsidies and incentives in various countries include SolarCity, a US-based solar panel installation company that has received tax incentives, rebates, and other subsidies in various US states; Vestas, a Danish wind turbine manufacturer that has received subsidies and tax breaks in a number of European countries, including Germany and Denmark; Tesla A US-based company that produces electric vehicles and has also developed energy storage products and solar panels The company has benefited from tax credits and other incentives in the US and other countries. Enel Green Power is an Italian energy company that has invested heavily in renewable energy, including wind, solar, and geothermal. The company has received subsidies and incentives in Italy and other countries, including China. General Nuclear Power Corporation (CGN) is a Chinese state-owned company that has been a major investor in renewable energy projects, including wind and solar; the company has benefited from government subsidies and other incentives in China.

It is difficult to guarantee that companies that receive subsidies and incentives for their renewable energy projects will not use those funds to support their operations in the fossil fuel industry, in spite of specific regulations and policies in place to prevent this from happening. For example, in the United States, companies receiving federal tax credits for renewable energy projects are required to document that the funds are being used specifically for those projects and not for other purposes. The Internal Revenue Service (IRS) also conducts audits to ensure that companies comply with these requirements. Similarly, in Europe, companies that receive subsidies and incentives for renewable energy projects are often subject to strict reporting and verification requirements to ensure that the funds are being used as intended.

It's worth noting, however, that these regulations and policies are not foolproof, and there is always the potential for companies to misuse subsidies and incentives. In addition, some critics argue that even if companies are using these funds as intended, they may be using them to prop up renewable energy projects that would not be economically viable without the subsidies rather than investing in more innovative and cost-effective technologies. Ultimately, it is up to policymakers and regulators to ensure that subsidies and incentives for renewable energy are being used effectively and efficiently to support the transition to a low-carbon economy.

The Texas Chapter 313 incentives, also known as the Texas Economic Development Act, provide a tax incentive programme for companies that invest in certain areas of the state, including renewable energy projects. Under this programme, eligible companies can receive a 10-year tax abatement on property taxes for a new project, which can significantly reduce the company's operating costs.

Many companies in the renewable energy sector have applied for Chapter 313 incentives in Texas, including major players like E.ON, NextEra Energy, and Invenergy. These companies are involved in a range of renewable energy fields, including wind power, solar power, and energy storage.

In order to promote renewable energy like solar and wind, the federal government of the US provides two types of incentives: the Production Tax Credit (PTC) and the Investment Tax Credit (ITC). Both the PTC and ITC amount to about one-third of the cost of building and operating the facilities. This programme expires on December 31, 2022. These incentives were introduced in 1976. So far, over 100 billion dollars of subsidies have been disbursed through 2020.

The main goal of this scheme has been introduced in the hope that the cost of producing power will decrease. According to the U.S. Energy Information Administration (EIA), the cost of production from wind dropped from $2,335 per kW in 2010 to $1,319 per kW in 2020, a drop of 43%. Similarly, in the case of solar, it dropped from $7,297 to $1,319 per kW, showing a drop of 82% (Stacy, Tom, and George Taylor, 2000). Obviously, the PTC and ITC subsidies have lowered production costs substantially for renewable project developers. But according to a study by the Energy Policy Institute at the University of Chicago, the electricity rates across the country increased by 11% when the share of renewable energy increased by 1.8% and by 175 when the renewable share increased by 4.2% (Davis, L. W., & Gertler, P. J. (2018)).

Additionally, foreign corporations own a large portion of the wind and solar power installed in the US, and they also profit from the tax credits that these power projects produce. According to a study, it has been found that out of the $24.5 billion in PTC credits in 2017 and 2016, just 15 companies received three-quarters of those credits, and 42%, amounting to $8.2 billion of that total, went to seven overseas companies (Erickson, Angela C., 2000).

The majority of solar and wind projects are in Texas, US. It has been found that these are owned by energy or utility companies known for fossil fuel companies, including some that have aggressively opposed renewable energy and climate policies (Rawlins, A. (2022, August 5). Of the 191 wind and solar applications filed in 2022 for solar and wind projects in Texas, it is extremely hard to track the owners of these companies because most of them are structured as limited liability companies.

It is clear from the above figure that only 69 of the 191 are actually owned by renewable energy companies, while 191, or more than half, are from energy companies known for oil and gas or utilities with fossil fuel assets, including BP. It will be intriguing to observe what happens after the completion of these projects, as the current analysis only provides a brief overview of the companies involved at the implementation stage. In fact, nearly fifty per cent of the projects constructed between 2020 and 2021 were sold to other companies by 2022, either through acquisition or sale. The prevailing trend in these sales has been that renewable energy companies are selling their projects to energy and utility companies. For instance, BP acquired Alpin Sun, and Duke Energy bought Recurrent Energy. It's important to note that many entities that have benefited from government subsidies have parent companies with higher carbon emissions.

Renewable energy sources such as solar, wind, hydro, geothermal, and biomass have significant potential to become major sources of energy in the future. It is estimated that the amount of solar energy that reaches the Earth's surface in one hour is enough to meet the world's energy needs for an entire year. It is a commonly cited figure in the renewable energy industry. This estimate is based on the total amount of solar energy that reaches the Earth's surface, which is known as solar irradiance. According to the National Renewable Energy Laboratory (NREL), the solar irradiance on a clear day at sea level is approximately 1,000 watts per square metre (W/m2).

The cost of renewable energy technologies has been declining rapidly, making them more competitive with fossil fuels. Solar PV: According to the International Renewable Energy Agency (IRENA), the global weighted-average cost of electricity from new utility-scale solar PV has decreased by 82% between 2010 and 2020. In 2020, the global levelized cost of electricity (LCOE) for new solar PV projects was estimated to be around $0.06 per kilowatt-hour (kWh) for the most competitive projects, with some projects achieving costs as low as $0.02 per kWh.

The declining cost of producing renewable energy has been a major driving force behind the increasing popularity of clean energy sources. In recent years, the cost of producing energy from renewable sources such as solar, wind, and hydropower has decreased significantly, making them increasingly competitive with traditional fossil fuel-based sources. Technological advancements, economies of scale, and supportive government policies have all contributed to this cost reduction. As a result, renewable energy is now often the most cost-effective choice for new energy generation projects in many parts of the world, leading to a significant shift towards a cleaner and more sustainable energy future.

According to IRENA, the global weighted-average cost of electricity from new onshore wind has dropped by 39% between 2010 and 2020. This is just one example of the declining cost of renewable energy. In 2020, the global LCOE for new onshore wind projects was estimated at around $0.05 per kWh for the most competitive projects. The cost of offshore wind has also been declining rapidly in recent years. According to BloombergNEF, the global LCOE for offshore wind projects fell by 71% between 2010 and 2020, from $243 to $71 per MWh. In 2020, the lowest-cost offshore wind projects were estimated at around $44 per MWh. The cost of battery storage, which is crucial for integrating intermittent renewable energy sources into the grid, has also been declining rapidly. According to IRENA, the global weighted-average cost of utility-scale lithium-ion battery storage systems decreased by 89% between 2010 and 2020, from $2,152 per kilowatt-hour (kWh) to $240 per kWh.

Solar energy and ocean energy are two types of renewable energy sources that have gained increasing attention in recent years as we strive to reduce our dependence on fossil fuels.

8.1.1.1 Solar Energy

Solar power has emerged as a revolutionary solution in the pursuit of clean and sustainable energy sources. By harnessing the abundant and renewable energy of the sun, solar energy has the potential to transform our energy systems and mitigate the negative impacts of traditional fossil fuel-based power generation. This essay explores the advantages, challenges, and future prospects of solar energy, emphasising its role in paving the way towards a more sustainable future.

Solar energy is obtained by converting sunlight into electricity through photovoltaic (PV) cells or by utilising solar thermal technologies. PV cells, also known as solar panels, consist of semiconducting materials that generate an electric current when exposed to sunlight. These cells can be installed on rooftops and solar farms or integrated into various devices to generate clean electricity. Solar thermal technologies, on the other hand, capture the sun's heat to produce steam, which can power turbines and generate electricity or provide direct heat for applications like water and space heating.

The sun provides an inexhaustible energy source, supplying an ample amount of solar radiation. It ensures a sustainable and virtually limitless source of clean energy, reducing our reliance on finite fossil fuels. Technological advancements, economies of scale, and supportive policies have significantly reduced the cost of solar energy. It offers long-term cost savings through reduced energy bills and attractive financial incentives like net metering and government subsidies.

Solar energy represents a clean, abundant, and sustainable solution to our pressing energy needs. As we strive for a more sustainable future and address the challenges of climate change, solar power emerges as a key player in the transition to a low-carbon economy. Embracing solar energy can significantly reduce greenhouse gas emissions, enhance energy security, and create a cleaner and healthier planet for future generations. With continued advancements and support, solar energy has the potential to revolutionise the way we produce and consume electricity, paving the way for a brighter, more sustainable future.

8.1.1.2 Ocean Energy

Ocean energy has significant potential as a renewable energy source, as it is abundant, predictable, and relatively high energy density. There are several types of ocean energy technologies that can be used to generate electricity, including tidal energy, wave energy, and ocean thermal energy conversion (OTEC).

Tidal energy is generated by harnessing the power of ocean tides as they rise and fall. This can be done using turbines, which are placed in tidal currents and use the flow of water to generate electricity. Wave energy, on the other hand, is generated by capturing the energy of ocean waves as they move towards the shore. This can be done using various technologies, including oscillating water columns and point absorbers.

OTEC is a technology that uses the temperature difference between warm surface waters and cold deep waters to generate electricity. This can be done using a closed-cycle or open-cycle system, depending on the specific technology used.

Despite its potential, ocean energy is still in the early stages of development and deployment, and there are several challenges that must be overcome in order to realise its full potential. One of the main challenges is the high cost of ocean energy technologies, particularly in the early stages of development. However, as the technology continues to mature and economies of scale are achieved, the cost is expected to decrease.

Another challenge is the limited availability of suitable sites for ocean energy projects, particularly tidal and wave energy, which require specific ocean conditions. There are also concerns about the environmental impact of ocean energy technologies, particularly on marine ecosystems and wildlife.

Despite these challenges, there is growing interest in ocean energy as a renewable energy source, and many governments and companies around the world are investing in research and development to advance the technology and bring down costs. The potential applications for ocean energy are vast, and it is likely to play an increasingly important role in the transition to a more sustainable energy system in the coming decades.

The world's oceans hold vast potential as a source of clean, renewable energy. Ocean energy technologies harness the power of waves, tides, and ocean currents to generate electricity, and they have the potential to play a significant role in reducing our reliance on fossil fuels. These technologies come in various forms, including wave energy converters, tidal turbines, and ocean current turbines, each designed to convert the kinetic energy of the ocean into usable electrical energy. While ocean energy is still a relatively nascent industry, significant progress has been made in developing and deploying these technologies in recent years. With continued innovation and investment, ocean energy has the potential to become a key component of our global energy mix, providing clean, reliable, and sustainable power for generations to come.

There are several ocean energy technologies available today, and ongoing research and development are leading to the emergence of new and improved technologies. Here are some of the updated ocean energy technologies: Wave energy converters that capture the kinetic energy of ocean waves and convert it into electricity. There are several types of wave energy converters, including point absorbers, oscillating water columns, overtopping devices, and tidal energy converters that harness the kinetic energy of ocean tides using underwater turbines. Several types of tidal energy converters, including horizontal axis turbines, vertical axis turbines, tidal fences, and ocean current energy converters, extract energy from ocean currents using underwater turbines. The most common type of ocean current energy converter is the horizontal axis turbine. Ocean thermal energy conversion (OTEC) is a technology that uses the temperature difference between warm surface waters and cold deep waters to generate electricity. OTEC systems typically use a closed-loop or open-loop cycle to transfer heat and generate power, and salinity gradient power technology involves harnessing the energy produced by the difference in salinity between freshwater and saltwater. There are various methods to capture energy from the difference in salt concentration in water, including reverse electrodialysis, pressure-retarded osmosis, and capacitive mixing. In addition to these techniques, floating offshore wind turbines have become popular in coastal areas, despite not being classified as ocean energy technology. These turbines are mounted on platforms that float on the water's surface and have the advantage of being able to operate in deeper waters compared to their fixed-bottom counterparts.

Despite its significant potential, ocean energy has not been fully utilised due to several technical, economic, and environmental challenges. Here are some key reasons: The high cost of building and deploying ocean energy technologies is a significant barrier to their widespread adoption. The development of new technologies and the high initial capital investment required to instal ocean energy systems remain significant challenges. Lack of infrastructure that is required for ocean energy systems require specialised infrastructure, such as offshore power cables and subsea connection equipment, which can be expensive to instal and maintain;

Despite these challenges, the development of ocean energy technologies is continuing, and there are ongoing efforts to overcome these barriers. As research and development efforts continue, it is likely that ocean energy will play an increasingly significant role in the global energy mix, helping to reduce carbon emissions and mitigate climate change.

Elizabeth C. Brown (Brown, E. C. 2018) provides a comprehensive overview of ocean energy technologies, their resource potential, and the economic, environmental, and regulatory factors that influence their deployment. Lynett and Edge (Lynett, P., & Edge, B. L. 2019) provide an overview of wave energy technologies, including the physical principles of wave energy conversion and the design and performance of wave energy converters. Nunes, P.A.L.D. (Nunes, P.A.L.D. 2018) and Abbasi and Bard (Abbasi, B., and Bard, J. 2018) provide a comprehensive overview of marine renewable energy, including wave, tidal, ocean current, and ocean thermal energy conversion technologies. It also discusses the potential environmental and social impacts of marine renewable energy deployment.

Tidal power is a form of renewable energy that harnesses the power of the tides to generate electricity. It works by capturing the energy of the rise and fall of ocean tides and converting it into electrical power.

There are two main types of tidal power systems, the tidal range and the tidal stream. Tidal range systems rely on the difference in height between high and low tides and typically use dams or barrages to capture the energy of the tide as it flows in and out.

On the other hand, tidal stream systems use underwater turbines that turn in response to the flow of water in tidal currents.

Tidal power has several advantages as a renewable energy source. Unlike wind and solar power, tidal power is predictable and can be relied upon to generate electricity at specific times of day. Additionally, tidal power systems have a relatively small environmental footprint and do not produce greenhouse gas emissions or other pollutants.

Building a tidal power plant can be expensive, as Tidal power plants require specific conditions to be effective, including strong tidal currents and large tidal ranges. These conditions are only present in a limited number of locations, which can limit the potential for widespread adoption of tidal power.

The construction of tidal power plants can have negative impacts on the surrounding marine environment, including disruption to wildlife and altered sediment patterns. These impacts must be carefully considered and managed in order to minimise harm to ecosystems.

Tidal power must compete with other renewable energy sources, such as wind and solar power, which have become increasingly cost-competitive in recent years. This competition can make it difficult for tidal power to attract investment and government support.

While tidal power has the potential to provide a significant source of renewable energy, it faces significant challenges that have prevented it from becoming a major focus of the renewable energy industry.

In summary, ocean energy has significant potential as a renewable energy source due to its abundance, predictability, and high energy density. Tidal energy, wave energy, and ocean thermal energy conversion (OTEC) are among the various technologies used to harness ocean energy. However, ocean energy is still in the early stages of development, and there are challenges to overcome, such as high costs, limited suitable sites, and potential environmental impacts. Nevertheless, there is growing interest in ocean energy, and governments and companies are investing in research and development to advance the technology. With continued innovation and investment, ocean energy has the potential to become an important component of our global energy mix and contribute to a more sustainable future.

8.2 Future of the Human Race

Around 200,000 years ago, humans emerged on Earth, and currently, there are approximately eight billion people inhabiting the planet. Numerous factors, such as technological advancements and difficulties presented by the natural environment, have had an impact on the course of human history. The pace of scientific advancement, the effects of climate change, and the rise in population will have a significant impact on the course that humanity will take in the decades and centuries to come. The future of the human race is a complex and multifaceted subject, encompassing aspects like the effects of climate change, the progress of technology, and the growth of the human population. Despite the significant obstacles that lie ahead, there are also opportunities for advancement and innovation that could result in a more sustainable and prosperous future for everyone.

The expansion of the world's population will also have a big influence on the path that the human race takes in the years to come. The United Nations has released research predicting that the global population will reach 9.7 billion by 2050. The majority of the population expansion is expected to take place in emerging countries. This population growth will create extra constraints on resources such as food, water, and energy, which has the potential to exacerbate already existing societal conflicts and inequities.

The issue of climate change is a crucial one that will have a significant impact on the development of the human race in the years to come. The development of renewable energy sources, such as solar and wind power, has the potential to reduce our dependence on fossil fuels and mitigate the negative effects of climate change. Furthermore, technological progress in fields such as biotechnology and medicine has the potential to bring about significant improvements in human health as well as increases in average longevity.

Compared to the brains of most other species, the human brain is considerably larger. On average, the adult human brain weighs around 1.3 kilogrammes. However, when considering the body sizes of different animals, the human brain is not the largest. There are various creatures, such as elephants, whales, and dolphins, that possess brains larger than ours.

It's worth noting that advancements in artificial intelligence (AI) and neuroscientific research have the potential to enhance our understanding of brain function and cognitive processes. These fields have made significant progress in recent years. Nonetheless, the future implications concerning changes in brain size or capabilities will primarily hinge on the specific advancements and their applications. Currently, it is challenging to predict with certainty what these advancements will entail and how they will be utilised.

The development of artificial intelligence (AI) is anticipated to play a significant role in determining the course that the human race will take in the future. The continued development of artificial intelligence has the potential to revolutionise practically every facet of human life, from healthcare and education to transportation and entertainment. This includes the possibility of creating entirely new professions. While artificial intelligence has the capacity to create major benefits, such as greater efficiency and productivity, it also raises concerns about its possible negative influence on jobs, privacy, and security.

Recent developments in artificial intelligence have made it possible for machines to execute more difficult jobs and digest information more quickly. However, integrating AI directly into the human brain is an attempt that is fraught with a great deal of difficulty. The brain is an organ that is extraordinarily complex and that we do not yet have a complete understanding of. Because of this, developing seamless communication between artificial and biological systems presents substantial difficulty.

Brain-computer interfaces, often known as BCIs, are one area of research that attempts to create a communication channel between the human brain and many other technologies that are external to the body. The use of brain-computer interfaces (BCIs) has demonstrated significant potential for enabling people to operate external devices, such as prosthetic limbs or computer interfaces, using only their thoughts. On the other hand, the amount of integration necessary for a cybernetic organism to be completely functional calls for a considerably higher level of complexity.

The combination of artificial intelligence (AI) with the human brain has the potential to result in the creation of a cybernetic creature, more commonly referred to as a cyborg. This is a possibility. The building of a fully integrated cybernetic organism is currently mostly theoretical and confronts a vast number of technical, ethical, and practical obstacles. While great progress has been achieved in the disciplines of artificial intelligence and brain-computer interfaces (BCIs), this endeavour is still in its infancy.

Henry Gee, a palaeontologist (Gee, H., 2021), proposes that mammal species have a tendency to have a brief existence, appearing, thriving, and then disappearing after a million years or so. Henry Gee continues by stating that the process of extinction is typically a gradual reaction to habitat destruction. The at-risk species, in this case, humans, have a tendency to colonise specific habitat patches to the detriment of other species, which then move on to other areas and become more dispersed. At this time, Homo sapiens is responsible for the consumption of anywhere between 25 and 40 per cent of the net primary productivity. The term "net primary productivity" refers to the organic material that plants create with the help of sunlight, water, and air. Millions of other species suffer as a result of a lack of resources.

The number of live births may be negatively impacted by the fact that human sperm quality has significantly decreased over the past few decades. The United Nations predicts that global fertility rates will continue to decline over the upcoming decades, which will cause the world population to gradually decline once it reaches its peak. It is feasible that the decrease in mortality rates expected as a result of improvements in healthcare and living circumstances would somewhat counterbalance the population loss that is brought on by lowering fertility rates. This would be the case if living conditions and healthcare were to improve. Although there is growing evidence to suggest that over the past few decades, the general quality of human sperm has been declining, predicting the trajectory of this trend into the future is challenging and dangerous.

Sperm quality indicators like sperm count, sperm motility, and sperm morphology (shape) have all decreased, according to a number of studies. Numerous factors, including but not limited to increased exposure to environmental toxins, sedentary lifestyles, poor diets, obesity, stress, and specific medical conditions, may be to blame for this decline. Factors such as these have been proposed as potential culprits. On the other hand, it is essential to keep in mind that not all studies come to the same conclusion about the scope and scale of the decrease, and there are other studies that find no substantial change.

Is it possible that there will be no trace of the human race?

There is a possibility that Homo sapiens, like all other species on Earth, could one day become extinct. Even though the likelihood of humans being extinct in the not-too-distant future is now quite low, there are a number of potential causes that could enhance this danger, such as the following:

In 2050, according to projections made by the United Nations, the global population is expected to reach a maximum of approximately 9.7 billion before beginning a slow but steady drop. However, the precise course that population trends will take beyond the year 2050 will depend on various factors, such as fertility rates, death rates, and migration patterns, all of which are notoriously difficult to accurately forecast. Although the intricate interactions of genetic, environmental, and cultural influences have not altered the rate or course of Homo sapiens' biological evolution, they have had an impact. The process of evolution remains continuous, and the human population is still subject to the influence of natural selection.

However, in recent times, the impact of natural selection on human populations has been modified as a result of advancements in medicine, nutrition, and other areas of modern life. These advancements have lowered the selective pressure from diseases and other environmental factors, which in turn has impacted the impact of natural selection on human populations. It is difficult to foresee with absolute precision the path that the evolution of humans will take in the future. Because human populations are so diverse and dynamic, a wide range of processes, such as genetic drift, gene flow, and natural selection, to name a few, have an impact on the evolution of these groups. Some researchers believe that the human population is always progressing. Even though humans have the largest brain size of any mammal, our ability to create and adapt has enabled us to successfully colonise almost every part of the world.

Recent advancements in genetic engineering, biotechnology, and artificial technology have the potential to have a large impact on the future of the human race in various ways. These possible effects can be broken down into two categories: positive and negative. The following are some possible repercussions that could result from these developments.

Genetic engineering and biotechnology are already being utilised to discover new therapies and cures for a variety of ailments, including cancer, genetic disorders, and infectious diseases. Some of the diseases being targeted include cancer, genetic disorders, and infectious diseases. It is very conceivable that we will observe an even greater number of ground-breaking achievements in the fields of medicine and healthcare as technology continues to improve.

It is not out of the question that humans could one day have better cognitive ability, increased physical strength, and other characteristics thanks to developments in technology such as brain-computer interfaces and genetic editing. However, these improvements can also result in the emergence of new types of inequality and prejudice.

The application of genetic engineering and biotechnology gives rise to a wide variety of ethical concerns on matters such as personal privacy, informed consent, and the inalienable rights of humans. Concerns have been raised over the possibility of autonomous weaponry as well as the influence that the development of artificial intelligence will have on employment opportunities and societal systems.

Biotechnology has the potential to assist in resolving environmental issues such as climate change and the depletion of natural resources. For instance, the cultivation of genetically modified crops might make it possible to boost food output while simultaneously lowering the amount of chemical pesticides and fertilisers required.

It is feasible that in the future, humans will be able to direct their own evolutionary process if they have the ability to change the genetic code of other animals. This gives rise to concerns regarding the potential ethical repercussions of such interventions as well as the effects that they could have on future generations.

Recent advancements in genetic engineering, biotechnology, and artificial technology have enormous implications for the future of the human race. In spite of the fact that these advancements present promising prospects for future medical and technological progress, substantial ethical and societal problems have been brought to light as a result of them, and these questions require careful consideration. It is critical that we approach these technologies with extreme caution and take into account the myriad of potential repercussions they could have not only for individuals but also for society as a whole.

In spite of all of these difficulties, there are good reasons to have hope for the future of the human species. We can make strides towards a more promising future for ourselves as well as for the generations that will come after us if we maintain our commitment to research and development and work together to find global solutions.

According to historian and author Yuval Noah Harari (Harari, Y.N. 2016), the future of humanity is influenced by the integration of biotechnology and artificial intelligence. Harari proposes that this combination could result in the emergence of a new kind of "superhuman" that surpasses regular humans in abilities and potentially establishes new social hierarchies if these two fields collaborate. This suggests a potential shift in the dominance of humans as the primary species on Earth due to technological advancements. In parallel, the future of energy is being shaped by various factors, including technological advancements, policy changes, and the recognition of sustainable energy sources. Renewable energy, such as solar, wind, hydroelectric, and geothermal power, is experiencing significant growth, aided by declining costs and increased investment in renewable infrastructure. Energy storage technologies are being developed to manage the intermittent nature of renewables, while smart grid technology is enhancing energy distribution efficiency. However, despite these advancements, global primary energy demand is projected to increase due to factors like population growth, rising incomes, and urbanisation in developing countries, as indicated by the IEA's World Energy Outlook 2021 report.

Endnotes

1. Abatzoglou, J., et al. (2007). A primer on global climate change and its likely impacts. In J.F.C. DiMento and P. Doughman (Eds.), Climate change: what it means for us, our children, and our grandchildren (pp. 11–44). Cambridge, MA: The MIT Press.

2. Abbasi, B., & Bard, J. (2018). Tidal Energy: Technology and Resource Characterization. John Wiley & Sons.

3. Andres, R.J., Marland, G., Boden, T., & Bischoff, S. (2000). Carbon dioxide emissions from fossil fuel consumption and cement manufacture, 1751–1991, and an estimate of their isotopic composition and latitudinal distribution. In T.M.L. Wigley & D.S. Schimel (Eds.), The Carbon Cycle (pp. 53–62). Cambridge University Press.

4. Anon, (2000) Analysis of carbon dioxide and other atmospheric gases. Australasian Institute of Geologists, News 86, 10–11

5. Bast, Joseph. (2010) 'Seven Theories of Climate Change.' Heartland Institute, 1 Jan. 2010.

6. Boden, T., Andres, R., & Marland, G. (2017). Global, Regional, and National Fossil-Fuel CO2 Emissions (1751– 2014) (V. 2017) [Data set]. Environmental System Science Data Infrastructure for a Virtual Ecosystem; Carbon Dioxide Information Analysis Center (CDIAC), Oak Ridge National Laboratory (ORNL), Oak Ridge, TN (United States). https://doi.org/10.3334/CDIAC/00001_V2017

7. Bošnjaković, Branko, (2010) After Copenhagen: Climate, Energy and Geopolitics // Energy and the Environment 2010, Engineering for a Low-Carbon Future, Vol I / Bernard Franković (ed.). Rijeka: Hrvatski savez za sunčevu energiju, 2010. pp. 1–35 (lecture, international peer review, full paper, scholarly)

8. Boute, A., & Zhikharev, A. (2019). Vested interests as driver of the clean energy transition: Evidence from Russia's solar energy policy. Energy Policy, 133, 110910. https://doi.org/10.1016/j.enpol.2019.110910

9. Brown, E. C. (2018). Ocean Energy: Technology and Resource Potential. Elsevier.

10. Carter, R.M., (2007, January 1). The myth of dangerous human-caused climate change. ResearchOnline@JCU. https://researchonline.jcu.edu.au/3130/

11. Cearreta, A. (2022). The Anthropocene perspective: A geological approach to climate change. Mètode: Revista de Difusió de la Investigació, (12).

12. Chakrabarty, D. (2009b). The climate of history: Four theses. Critical Inquiry, 35(2), 197–222. https://doi.org/10.1086/596640

13. Chaturvedi, Sanjay, and Timothy Doyle (2016). Climate Terror: A Critical Geopolitics of Climate Change. Springer, 2016.

14. Chumakov, N.M. (2004). Trends in Global Climate Changes Inferred from Geological Data. Stratigraphy and Geological Correlation. 12, 117–138.

15. Cohen, A.S., Coe, A.L., & Kemp, D.B. (2007). The Late Palaeocene Early Eocene and Toarcian (Early Jurassic) carbon isotope excursions: a comparison of their time scales, associated environmental changes, causes and consequences. Journal of the Geological Society, 164, 1093-1108.

16. Cook, J., Nuccitelli, D., Green, S. A., Richardson, M., Winkler, B., Painting, R., Way, R., Jacobs, P., & Skuce, A. (2013). Quantifying the consensus on anthropogenic global warming in the scientific literature. Environmental Research Letters, 8(2), 024024. https://doi.org/10.1088/1748-9326/8/2/024024

17. Crichton, M. (2004). State of fear. HarperCollins.

18. Crocker, T. D. (1966). The structuring of atmospheric pollution control systems. The economics of air pollution, 61, 81–84.

19. Crutzen, P. J., & Stoermer, E. F. (2000). The "Anthropocene". Global Change International Geosphere-Biosphere Programme Newsletter, 41, 17–18

20. Curry, J. A., & Webster, P. J. (2011). Climate science and the uncertainty monster. Bulletin of the American Meteorological Society, 92(12), 1667–1682. https://doi.org/10.1175/2011bams3139.1

21. Davis, L. W., & Gertler, P. J. (2018). The Economic Impacts of Electricity Deregulation: Evidence from the United States. Energy Journal, 39(2), 25–47. doi:10.5547/01956574.39.2.ldav.

22. Desai, Nitin. (2012) "The Geopolitics of Climate Change." Handbook of Climate Change and India, edited by Navroj Dubash, Routledge, 2012, http://dx.doi.org/10.4324/9780203153284.ch28.

23. Ding, J., Chen, J., & Tang, W. (2022). *Increasing trend of Precipitable Water Vapour in Antarctica and Greenland*. Copernicus GmbH. http://dx.doi.org/10.5194/egusphere-egu22-6891

24. Doran, P. T., & Kendall Zimmerman, M. (2009). Examining the Scientific Consensus on Climate Change. Eos, Transactions American Geophysical Union, 90(3), 22. doi:10.1029/2009eo030002.

25. Eamus, D. (1996). Responses of field-grown trees to CO_2 enrichment. The Commonwealth Forestry Review, 75(1), 39–47. https://doi.org/10.2307/42607274

26. Eddy, J. A. (1983). The Maunder Minimum: A reappraisal. Solar Physics, 89(1), 195–207. https://doi.org/10.1007/bf00211962

27. Erickson, A. C. (2018). Texas Public Policy Foundation. (2019, January 29). The Production Tax Credit: Corporate Subsidies & Renewable Energy.https://www.texaspolicy.com/the-production-tax-credit-corporate-subsidies-renewable-energy/

28. Fearnow, B. (2019, September 17). Psychologists warn parents and climate change alarmists against causing "eco-anxiety" in children. Newsweek.https://www.newsweek.com/eco-anxiety-climate-change-parent-fear-discussion-children-global-warming-depression-effects-1459731

29. Fischer, H., Wahlen, M., Smith, J., Mastroianni, D., & Deck, B. (1999). Ice core records of atmospheric CO2 around the last three glacial terminations. Science, 283(5408), 1712–1714. https://doi.org/10.1126/science.283.5408.1712

30. Friedlingstein, P., Jones, M. W., O'Sullivan, M., Andrew, R. M., Hauck, J., Peters, G. P., Peters, W., Pongratz, J., Sitch, S., Quéré, C. L., Bakker, D. C. E., Canadell, J. G., Ciais, P., Jackson, R. B., Anthoni, P., Barbero, L., Bastos, A., Bastrikov, V., Becker, M., ... Zaehle, S. (2019). Global carbon budget 2019. Earth System Science Data, 11(4), 1783-1838. https://doi.org/https://doi.org/10.5194/essd-11-1783-2019

31. Friis-Christensen, E. (1993). Solar activity variations and global temperature. Energy, 18(12), 1273–1284. https://doi.org/10.1016/0360-5442(93)90015-6

32. Gee, H. (2021, November 30). Humans are doomed to go extinct. ScientificAmerican.https://www.scientificamerican.com/article/humans-are-doomed-to-go-extinct/

33. Gray, William Mulroy, (2009). "Climate Change: Driven by the Ocean, not Human Activity." (2009).

34. Hamilton, Clive. (2012)'Theories of Climate Change.' Australian Journal of Political Science, vol. 47, no. 4, Dec. 2012, pp. 721–29, https://doi.org/10.1080/10361146.2012.732213.

35. Hansen, J., Sato, M., Hearty, P., Ruedy, R., Kelley, M., Masson-Delmotte, V., ... Zachos, J. C. (2016). Ice melt, sea level rise and superstorms: Evidence from paleoclimate data, climate modelling, and modern observations that 2 °C global warming could be dangerous. Atmospheric Chemistry and Physics, 16(6), 3761-3812. doi: 10.5194/acp-16-3761-2016

36. Harari, Y.N. (2016). "Yuval Noah Harari on the Future of Humanity" in The Guardian. Retrieved from: https://www.theguardian.com/books/2016/sep/15/yuval-noah-harari-on-the-future-of-humanity-sapiens-sequel-homo-deus

37. Houghton, J. (2009). Global warming: The complete briefing (4th ed.). Cambridge University Press.

38. Huang, J.-G., Bergeron, Y., Denneler, B., Berninger, F., & Tardif, J. (2007c). Response of forest trees to increased atmospheric carbon dioxide. Critical Reviews in Plant Sciences, 26(5–6), 265–283. https://doi.org/10.1080/07352680701626978

39. Humlum, O. (2022). The State of the Climate in 2021. Global Warming Policy Foundation, 51.

40. Hunt, E. (2021). The Trick review: How the Climategate scandal rocked the World. New Scientist. https://www.newscientist.com/article/2294061-the-trick-review-how-the-climategate-scandal-rocked-the-world/

41. Inhofe, J. M. (2012). The Greatest Hoax: How the Global Warming Conspiracy Threatens Your Future. Wnd Books.

42. Jaworowski, Z. (2007). Carbon dioxide: The Greatest Scientific Scandal of Our Time. Executive Intelligence Review, 34(11), 38-53.

43. Joos, F., Roth, R., Fuglestvedt, J. S., Peters, G. P., Enting, I. G., von Bloh, W., Brovkin, V., Burke, E. J., Eby, M., Edwards, N. R., Friedrich, T., Frölicher, T. L., Halloran, P. R., Holden, P. B., Jones, C., Kleinen, T., Mackenzie, F. T., Matsumoto, K., Meinshausen, M., ... Weaver, A. J. (2013). ACP - Carbon dioxide and climate impulse response functions for the computation of greenhouse gas metrics: A multi-model analysis. Atmospheric Chemistry and Physics, 13(5), 2793-2825. https://doi.org/https://doi.org/10.5194/acp-13-2793-2013

44. Kleidon, A., Messori, G., Baidya Roy, S., Didenkulova, I., & Zeng, N. (2023). Editorial: Global warming is due to an enhanced greenhouse effect, and anthropogenic heat emissions currently play a negligible role at the global scale. Earth System Dynamics, 14(1), 241–242. https://doi.org/10.5194/esd-14-241-2023

45. Lear, C. H., Anagnostou, E., Armstrong McKay, D., Cobb, K. M., Elderfield, H., Friedrich, O., ... Wilson, P. A. (2020). Geological Society

of London scientific statement: What the geological record tells us about our present and future climate. Journal of the Geological Society, 178(1), jgs2020-239. https://doi.org/10.1144/jgs2020-239

46. Lear, C. H., et al. (2020). Geological Society of London Scientific Statement: What the Geological Record Tells Us about Our Present and Future Climate. Journal of the Geological Society, 178(1). https://doi.org/10.1144/jgs2020-239.

47. Lee, X., & Fung, I. (2008). Decreased river runoff in central Asia during the late 20th century: evidence from Lake Balkhash. Journal of Climate, 21(15), 3952–3966.

48. Lewandowsky, S., and J. Cook. (2020). The Conspiracy Theory Handbook. Center for Climate Change Communication | George Mason University Foundation. https://www.climatechangecommunication.org/conspiracy-theory-handbook/.

49. Lindzen, Richard S., et al. (2001) "Does the Earth Have an Adaptive Infrared Iris?" Bulletin of the American Meteorological Society, vol. 82, no. 3, Mar. 2001, pp. 417–32, https://doi.org/10.1175/1520-0477(2001)082<0417:dtehaa>2.3.co;2.

50. Lomborg, B. (2020, July 11). How climate change alarmists are actually endangering the planet. New York Post. https://nypost.com/2020/07/11/how-climate-change-alarmists-are-actually-damaging-the-planet/

51. Lowenstein, T. K., & Demicco, R. V. (2006). Elevated eocene atmospheric carbon dioxide and its subsequent decline. Science, 313(5795).

52. Lynett, P., & Edge, B. L. (2019). Ocean Energy Systems: Wave Energy. Springer.

53. MacFarling Meure, C., Etheridge, D., Trudinger, C., Steele, P., Langenfelds, R., van Ommen, T., Smith, A., & Elkins, J. (2006). Law Dome carbon dioxide, CH4 and N2O ice core records extend to 2000 years BP. Geophysical Research Letters, 33(14), L14810.

54. Mann, G. (2018). Climate Leviathan: A Political Theory of Our Planetary Future. Verso Books.

55. McQueen, A. (2021). The Wages of Fear? In Philosophy and Climate Change (pp. 152–177). Oxford University Press. https://doi.org/10.1093/oso/9780198796282.003.0008

56. Metz, B., Davidson, O., et al. (2007). Climate Change 2007 – Mitigation of Climate Change. Contribution of Working Group III to the 4th Assessment of the IPCC. Cambridge University Press.

57. Meyers, P. A., & Lallier-Vergès, E. (1999). Lacustrine sedimentary organic matter records of Late Quaternary paleoclimates. Journal of Paleolimnology, 21(3), 345–372.

58. Montford, A. W. (2010). *The hockey stick illusion: Climategate and the corruption of science*. Stacey International Publishers.

59. Myers, Krista F., et al. (2021) "Consensus Revisited: Quantifying Scientific Agreement on Climate Change and Climate Expertise among Earth Scientists 10 Years Later." Environmental Research Letters, vol. 16, no. 10, Oct. 2021, p. 104030, doi:10.1088/1748-9326/ac2774.

60. Nunes, P. A. L. D. (2018). Marine Renewable Energy: Resource Characterization and Physical Effects. Springer.

61. Oreskes, N. (2004). The Scientific Consensus on Climate Change. Science, 306(5702), 1686-1686. doi:10.1126/science.1103618

62. O'Neill, S., & Nicholson-Cole, S. (2009). 'Fear won't do it.' Science Communication, 30(3), 355–379. https://doi.org/10.1177/1075547008329201

63. Panofsky, Hans A. (1956). "Theories of Climate Change." Weatherwise 9 (1956): 183-204.

64. Panofsky, Hans A. (1956) "Theories of Climate Change." Weatherwise 9 (1956): 183-204.

65. Payne, Rodger A. (2007) "Climate Change, Spring/Summer 2007, Issue 16." Sustain Magazine, vol. 2007, no. 16, Sept. 2019, doi:10.55504/2689-7296.1035.

66. Petit, J., Jouzel, J., Raynaud, D., Barkov, N. I., Barnola, J. M., Basile, I., ... & Stievenard, M. (1999). Climate and atmospheric history of the past 420,000 years from the Vostok ice core, Antarctica. Nature, 399 (6735), 429-436. https://doi.org/10.1038/20859

67. Pielke Jr., R. (2005). Misdefining "Climate Change" and the Implications for Science and Policy. Political Science Quarterly, 120(4), 601-626

68. Plimer, I. (2009). Heaven and Earth: Global Warming, the Missing Science. Connor Court.

69. Quéré, C. L., Moriarty, R., Andrew, R. M., Peters, G. P., Ciais, P., Friedlingstein, P., Jones, S. D., Sitch, S., Tans, P., Arneth, A., Boden, T. A., Bopp, L., Bozec, Y., Canadell, J. G., Chini, L. P., Chevallier, F., Cosca, C. E., Harris, I., Hoppema, M., & Zeng, N. (2014). ESSD - Global carbon budget 2014. Earth System Science Data, 7(1), 47-85. https://doi.org/https://doi.org/10.5194/essd-7-47-2015.

70. Rawlins, A. (2022, August 5). Who, exactly, benefits from renewable energy subsidies? The answer will surprise you. Fast Company. https://www.fastcompany.com/90776050/who-exactly-benefits-from-renewable-energy-subsidies-the-answer-will-surprise-you

71. Reid, G. C. (1987). Influence of solar variability on global sea surface temperatures. Nature, 329(6135), 142–143. https://doi.org/10.1038/329142a0

72. Ross, R. (1999). Https://culteducation.com/. Https://Culteducation. Com/. https://culteducation.com/warningsigns.html

73. Saxe, H., Ellseworth D., & Heath, J. (1998). Trees and forests functioning in an enriched carbon dioxide atmosphere. The New Phytologist, 139(3), 395-436. doi:10.1046/j.1469-8137.1998.00221.x

74. Schimmelmann, A., Albert, D. B., & Lyons, T. W. (2006). Carbon isotope geochemistry of coalbed methane: a global data review and new data from the Powder River Basin. International Journal of Coal Geology, 66(1-2), 22-44.

75. Soon, W., Baliunas, S., Idso, S. B., Kondratyev, K. Y., & Posmentier, E. S. (1999). Environmental effects of increased atmospheric carbon dioxide. Climate Research, 13(2), 149-164.https://www.int-res.com/abstracts/cr/v13/n2/p149-164/.

76. Spencer, Roy W., et al. (2007) "Cloud and Radiation Budget Changes Associated with Tropical Intraseasonal Oscillations." Geophysical Research Letters, vol. 34, no. 15, Aug. 2007, https://doi.org/10.1029/2007gl029698.

77. Stacy, T., & Taylor, G. (n.d.). The levelized cost of electricity from existing generation resources. Heartland Institute. Retrieved April 6, 2023, from https://www.heartland.org/publications-resources/publications/the-levelized-cost-of-electricity-from-existing-generation-resources

78. Sud, Y. C., et al. (1999) "Mechanisms Regulating Sea-Surface Temperatures and Deep Convection in the Tropics." Geophysical Research Letters, vol. 26, no. 8, Apr. 1999, pp. 1019–22, https://doi.org/10.1029/1999gl900197.

79. Sud, Y. C., et al. (1999) "Mechanisms Regulating Sea-Surface Temperatures and Deep Convection in the Tropics." Geophysical Research Letters, vol. 26, no. 8, Apr. 1999, pp. 1019–22, https://doi.org/10.1029/1999gl900197.

80. Summerhayes, C. P., DeConto, R. M., & Goddéris, Y. (2013). The geological perspective of global warming. Global Warming Policy Foundation. https://www.thegwpf.org/content/uploads/2013/12/Geological-Perspective.pdf

81. Svensmark, H. (1997). Variation of Cosmic Ray Flux and Global Cloud Coverage—A New Dawn for Solar Climate Relations? Journal of Aerosol Science, 28(6), 1102. https://doi.org/10.1016/s0021-8502(97)88096-2.

82. Syvitski, J., Waters, C., Day, J., Milliman, J. D., Summerhayes, C., Steffen, W., Zalasiewicz, J., Cearreta, A., Galuszka, A., Hajdas, I., Head, M. J.,

Leinfelder, R., McNeill, J. R., Poirier, C., Rose, N. L., Shotyk, W., Wagreich, M., & Williams, M. (2020). Extraordinary human energy consumption and resultant geological impacts beginning around 1950 CE initiated the proposed Anthropocene Epoch. Communications Earth & Environment, 1, 32. https://doi. org/10.1038/s43247-020-00029-y

83. Tyagi, A., & Carley, K. M. (2021). Climate Change Conspiracy Theories on Social Media. arXiv preprint arXiv:2107.03318.

84. Willis, J.R., & Dales, J.H. (1969). Pollution, Property, and Prices. The University of Toronto Law Journal, 19, 277.

85. World Resources Institute. (2010) Climate Analysis Indicators Tool (CAIT). Retrieved from http://cait.wri.org/

86. Wright, C., & Nyberg, D. (2015). Climate change, capitalism, and corporations: Processes of creative self-destruction. Cambridge University Press.

87. Wright, C., & Nyberg, D. (2015). Climate change, capitalism, and corporations: Processes of creative self-destruction. Cambridge University Press.

88. Young, Laura D., and Erin B. Fitz. (2021) "Who Are the 3 Per Cent? The Connections Among Climate Change Contrarians." British Journal of Political Science, vol. 52, no. 4, Dec. 2021, pp. 1503–22, doi:10.1017/s0007123421000442.

Bibliography

Alinejad, D., & Van Dijck, J. (2022, November). Climate Communication: How Researchers Navigate between Scientific Truth and Media Publics. Communication and the Public, 1-19. Advance online publication.

https://doi.org/10.1177/20570473221138612

Al-Zu'bi, Y., Adejuwon, J. A., Amusan, L., Anyamba, E., Araos, M., Barbier, E. B., ... & Mohamed-Katerere, J. C. (2022, November). African Perspectives on Climate Change Research. Nature Climate Change, 1-7.

https://doi.org/10.1038/s41558-022-01519-x.

American Association for the Advancement of Science. (2014). Climate change: How do we know? (Report). Retrieved from

https://www.aaas.org/climate-change-how-do-we-know

Andres, R. J., Marland, G., Boden, T., & Bischof, S. (2000). Carbon dioxide emissions from fossil fuel consumption and cement manufacture, 1751-1991, and an estimate of their isotopic composition and latitudinal distribution. In T.M.L. Wigley & D.S. Schimel (Eds.), The Carbon Cycle (pp. 53-62). Cambridge University Press.

Angus, I. (2016). Facing the Anthropocene: Fossil Capitalism and the Crisis of the Earth System. NYU Press.

ANU College of Science. (2019a). How we discovered the climate problem. Retrieved from

https://science.anu.edu.au/news-events/news/how-we-discovered-climate-problem

Archer, D., Eby, M., Brovkin, V., Ridgwell, A., Cao, L., Mikolajewicz, U., Caldeira, K., Matsumoto, K., Munhoven, G., Montenegro, A., & Tokos, K. (2009). Atmospheric Lifetime of Fossil Fuel Carbon Dioxide. Annual Review of Earth and Planetary Sciences, 37, 117-134.

https://doi.org/10.1146/annurev.earth.031208.100206

Archibald, D. (2014). Solar cycle 24: Implications for the United States. Journal of American Physicians and Surgeons, 19(1), 4-9.

Arnold, D. G. (2012). Climate change, catastrophism, and eco-philosophy. Environmental Philosophy, 9(1), 3-22. doi: 10.5840/envirophil2012921.

Asbi, A. M., & Siregar, D. I. (2022). Philosophy of science analysis in studies related to climate change impacts on the resilience of informal areas. Jurnal Perencanaan Dan Pengembangan Kebijakan, 2(3), 230. https://doi.org/10.35472/jppk.v2i3.854.

Attfield, R. (2015). Climate Change and Philosophy: Transformational Possibilities. Bloomsbury Academic.

Barcellos, A. L., Saccol, R. D. S. P. Carvalho, N. L. Rosa, L. F. (2019). A simple reflection on climate change. Revista Eletrônica Em Gestão, Educação e Tecnologia Ambiental, 23, 18. https://doi.org/10.5902/2236117034387.

Barnett, J. (2007). The geopolitics of climate change. Geography Compass, 1(6), 1361–1375. https://doi.org/10.1111/j.1749-8198.2007.00066.x

Bast, Joseph. 'Seven Theories of Climate Change.' Heartland Institute, 1 Jan. 2010, https://www.heartland.org/publications-resources/publications/7-theories-of-climate-change?source=policybot.

Bastardi, J. (2018). The Climate Chronicles: Inconvenient Revelations You Won't Hear from Al Gore--And Others. Createspace Independent Publishing Platform.

Bastardi, J. (2020). The Weaponization of Weather in the Phoney Climate War (J. Payne, Ed.). Gatekeeper Press.

Beardsworth, R. (2020). Climate science, the politics of climate change and futures of IR. International Relations, 34(3), 374-390. doi:10.1177/0047117820946365

Beck, E. G. (2009). The CO2 temperature relationship: fact or fiction? Energy and Environment, 20(8), 1049-1058. https://doi.org/10.1260/0958-305X.20.8.1049

Beck, U. (2010). Climate for Change, or How to Create a Green Modernity? Theory, Culture & Society, 27(2–3), 254–266. https://doi.org/10.1177/0263276409358729

Bell, L. (2011). Climate of Corruption: Politics and Power Behind the Global Warming Hoax. Greenleaf Book Group.

Berger, A., Dickenson, R. E., & Kidson, J. W. (1989). Understanding climate change. American Geophysical Union.

Blaauw, H. J. (2017). Global Warming: Sun and Water. Energy & Environment, 28(4), 468-483. https://doi.org/10.1177/0958305X17695276

Black, R. (2011, November 28). A brief history of climate change. BBC News. https://www.bbc.com/news/science-environment-15874560

Boardman, B., & Darby, S. (2013). The effectiveness of financial incentives for energy efficiency in the UK domestic sector: A review of the evidence. Energy Policy, 60, 7-21.

Bohr, J. (2016). The 'Climatism' Cartel: Why Climate Change Deniers Oppose Market-Based Mitigation Policy. Environmental Politics, 25(5), 812-830. doi:10.1080/09644016.2016.1156106.

Bolin, Bert. (2007). A History of the Science and Politics of Climate Change: The Role of the Intergovernmental Panel on Climate Change. Cambridge University Press.

Bosnjakovic, B. (2012). Geopolitics of Climate Change: A Review. Proceedings of the International Conference "Energy, Environment and Climate Change," 8-10 June 2012, Dubrovnik, Croatia, 187-198.

Bošnjaković, Branko, After Copenhagen: Climate, Energy and Geopolitics // Energy and the Environment 2010, Engineering for a Low-Carbon Future, Vol I / Bernard Franković (ed.). Rijeka: Hrvatski savez za sunčevu energiju, 2010. pp. 1-35 (lecture, international peer review, full paper, scholarly)

Boute, A., & Zhikharev, A. (2019). Vested interests as driver of the clean energy transition: Evidence from Russia's solar energy policy. Energy Policy, 133, 110910. doi: 10.1016/j.enpol.2019.110910.

Bowen, F., & Aragon-Correa, J. A. (2014). Greenwashing in Corporate Environmentalism Research and Practice: The Importance of What We Say and Do. Organization & Environment, 27(2), 107–112. https://doi.org/10.1177/1086026614537078

Brennan, A., & Lo, Y. Y. (Eds.). (2011). Climate change and global justice. Cambridge, MA: MIT Press.

Brigham-Grette, J., et al. (2006). Petroleum Geologists' award to novelist Crichton is inappropriate. Eos, Transactions American Geophysical Union, 87(36), 364-364. doi: 10.1029/2006EO360008.

Broecker, W. S. (2019). The catastrophic Anthropogenic Global Warming (CAGW) hypothesis and its testing. Reviews of Geophysics, 57(1), 173-184. https://doi.org/10.1029/2018RG000610

Brounen, D., Kok, N., & Quigley, J. M. (2014). Energy policy and the Green Deal. Journal of Environmental Economics and Management, 67(2), 175-191.

Buettner, A. (2010). Climate change in the media: Climate denial, Ian Plimer, and the staging of public debate. MEDIANZ: Media Studies Journal of Aotearoa New Zealand, 9(1), 79-97. doi: 10.11157/medianz-vol12iss1id48

Bureau, G. B. (2021, December 16). The seven sins of greenwashing. Green Business Bureau.

https://greenbusinessbureau.com/green-practices/the-seven-sins-of-greenwashing/

Calder, G. (2017). Climate change as a challenge to philosophy. In R. Giacchetti & J. Henderson (Eds.), Climate Change and the Humanities (pp. 159-175). Palgrave Macmillan.

https://doi.org/10.1057/978-1-137-55124-5_8

Callaghan, M.W., Minx, J.C., & Forster, P.M. (2020). A topography of climate change research. Nature Climate Change, 10, 118-123.

https://doi.org/10.1038/s41558-019-0679-9

Caney, S. (2010). Climate change and the future: Discounting time, wealth, and risk. Journal of Social Philosophy, 41(3), 323-342.

https://doi.org/10.1111/j.1467-9833.2010.01471.x

Carbon Dioxide Information Analysis Center. (n.d.). Carbon Dioxide Information Analysis Center (CDIAC) [Website]. Oak Ridge National Laboratory, US Department of Energy. Retrieved [date], from

http://cdiac.ornl.gov/

Carter, R. M. (2007). Counting the cost of climate change. Energy & Environment, 18(1), 1-12.

Carter, R. M. (2008). Climate crunch: The state of climate science. Global Research and Reporting.

Carter, R. M. (2009). Why I am sceptical about climate change. Daily Telegraph, 10.

Carter, R. M. (2012). Is climate change really dangerous? Bulletin of Canadian Petroleum Geology, 60(4), 323-327.

Carter, R. M. (2013). Taxing Air: Facts and fallacies about climate change. Connor Court Publishing.

Carter. (2007, January 1). The myth of dangerous human-caused climate change. ResearchOnline@JCU.

https://researchonline.jcu.edu.au/3130/

Cavicchioli, R., Ripple, W. J., Timmis, K. N., Azam, F., Bakken, L. R., Baylis, M., Behrenfeld, M. J., Boetius, A., Boyd, P. W., Classen, A. T., Crowther, T. W., Danovaro, R., Foreman, C. M., Huisman, J., Hutchins, D., Jansson, J. K., Karl, D. M., Koskella, B., Mark Welch, D. B., Martiny, J. B., Moran, M. A., Orphan, V. J., Reay, D., Remais, J. V., Rich, V. I., Singh, B. K., Stein, L. Y., Stewart, F. J., Sullivan, M. B., van Oppen, M. J., Weaver, S. C., Webb, E. A., & Webster, N. S. (2019). Scientists' warning to humanity:

Microorganisms and climate change. Nature Reviews Microbiology, 17, 569-586. doi: 10.1038/s41579-019-0222-5

Cearreta, A. (2022, January). The Anthropocene Perspective: A Geological Approach to Climate Change. Mètode Revista de Difusió de La Investigació, (12).

Chakrabarty, D. (2009). The climate of history: Four theses. Critical Inquiry, 35(2), 197–222.
https://doi.org/10.1086/596640

Chaturvedi, S., & Doyle, T. (2016). Climate terror: A critical geopolitics of climate change. Springer.

Chilcoat, C. (2014, November 5). Who stands to benefit from climate change. Oilprice.Com.
https://oilprice.com/The-Environment/Global-Warming/Who-Stands-To-Benefit-From-Climate-Change.html

Christy, J. R., & McKitrick, R. R. (2018). Overestimation of warming over the past 20 years. Atmospheric Science Letters, 19(2), e830.
https://doi.org/10.1002/asl.830

Chylek, P., Dubey, M. K., & Lesins, G. (2014). Tropical atmospheric temperature changes since 1979 have not been consistent with model predictions. Geophysical Research Letters, 41, 1-6.
https://doi.org/10.1002/2014GL060055

Cohen, A. S., Coe, A. L., & Kemp, D. B. (2007). The Late Palaeocene Early Eocene and Toarcian (Early Jurassic) carbon isotope excursions: a comparison of their time scales, associated environmental changes, causes and consequences. Journal of the Geological Society, 164, 1093-1108.

Cohen, S. (2020, August 24). Climate Change and the American Political Agenda. Columbia Climate School. https://news.climate.columbia.edu/2020/08/24/climate-change-american-political-agenda/.

Cook, J., Nuccitelli, D., Green, S. A., Richardson, M., Winkler, B., Painting, R., ... & Skuce, A. (2013). Quantifying the consensus on anthropogenic global warming in the scientific literature. Environmental Research Letters, 8(2).
https://doi.org/10.1088/1748-9326/8/2/024024

Cook, J., Nuccitelli, D., Green, S. A., Richardson, M., Winkler, B., Painting, R., ... Skuce, A. (2016). Consensus on consensus: a synthesis of consensus estimates on human-caused global warming. Environmental Research Letters, 11(4), 048002.
https://doi.org/10.1088/1748-9326/11/4/048002

Cowan, D., & Sherry-Brennan, F. (2014). Why the Green Deal failed. Energy Policy, 66, 393-405.

Crichton, M. (2004). State of Fear. HarperCollins.

Crutzen, P. J., & Stoermer, E. F. (2000). The "Anthropocene". Global Change International Geosphere-Biosphere Programme Newsletter, 41, 17–18.

Curry, J. A. (2008). A call for a reasoned approach to climate change. Journal of Policy Modeling, 30(3), 465-472. doi: 10.1016/j.jpolmod.2007.06.002

Curry, J. A. (2011). Climate change: Attribution and the scientific method. Science & Education, 20(5-6), 555-573.

https://doi.org/10.1007/s11191-011-9352-4

Curry, J. A. (2017). Climate sensitivity reconsidered. Atmospheric and Oceanic Physics, 53(5), 508- 515.

Curry, J. A. (2019). Climate change: The null hypothesis. Climate Dynamics, 53, 21–33.

https://doi.org/10.1007/s00382-019-04810-7

Dalby, S. (2013). The Geopolitics of Climate Change. Political Geography, 37, 38-47. doi:10.1016/j.polgeo.2013.09.004.

Desai, N. (2012). The Geopolitics of Climate Change. In N. Dubash (Ed.), Handbook of Climate Change and India (pp. 473-487). Routledge. doi: 10.4324/9780203153284.ch28

Doran, P. T., & Kendall Zimmerman, M. (2009). Examining the scientific consensus on climate change. Eos, Transactions American Geophysical Union, 90(3), 22.

Dörries, M. (2010). Climate catastrophes and fear. WIREs Climate Change, 1(6), 885-890.

https://doi.org/10.1002/wcc.79

Douglass, D. H., & Christy, J. R. (2008). The role of natural cycles in global warming. Theoretical and Applied Climatology, 95(1-2), 1-15.

Dryzek, J. S., & Pickering, J. (2018). The politics of the Anthropocene. Oxford University Press, USA..

Dunlap, R. E., & Jacques, P. J. (2013). Climate change denial books and conservative think tanks. American Behavioral Scientist, 57(6), 699-731.

https://doi.org/10.1177/0002764213477096.

Duraisamy, P., & Jérôme, A. (2017). Who wins in the Indian Parliament election: Criminals, wealthy and incumbents? Journal of Social and Economic Development, 19(2), 245-262. doi:10.1007/s40847-017-0044-0.

Earl, R. (2016, December 14). Fact-Checking The Claim Of 97% Consensus On Anthropogenic Climate Change. Forbes.

https://www.forbes.com/sites/uhenergy/2016/12/14/fact-checking-the-97-consensus-on-anthropogenic-climate-change/

Easterbrook, D. (2011). Evidence-Based Climate Science: Data Opposing Carbon Dioxide Emissions as the Primary Source of Global Warming. Elsevier.

Easterling, D. R., & Wehner, M. F. (2009). Is the climate warming or cooling? Geophysical Research Letters, 36(8), L08706.

Editor. (2019). The great climate conundrum. Nature Geoscience, 12(8), 581-581.

https://doi.org/10.1038/s41561-019-0428-1.

Edwards, M. (2017). The rise and fall of the theory of catastrophic anthropogenic global warming. The Journal of Forensic Sciences and Research, 1(1), 555554.

Emmett, R., & Lekan, T. (Eds.). (2016, March 2). Whose Anthropocene? Revisiting Dipesh Chakrabarty's "four theses." Environment & Society Portal.

https://www.environmentandsociety.org/perspectives/2016/2/whose-anthropocene-revisiting-dipesh-chakrabartys-four-theses

Emmott, S. (2013, June 29). Humans: The real threat to life on Earth. The Guardian.

https://www.theguardian.com/environment/2013/jun/30/stephen-emmott-ten-billion

Engle, R. F., Giglio, S., Lee, H., Kelly, B., & Stroebel, J. (2019). Hedging Climate Change News. Microeconomics: Decision-Making under Risk & Uncertainty eJournal.

Epstein, A. (2015, January 6). "97% of climate scientists agree" is 100% wrong. Forbes.

https://www.forbes.com/sites/alexepstein/2015/01/06/97-of-climate-scientists-agree-is-100-wrong/?sh=76bfd9803f9f

Erickson, Angela C. (2000). The Production Tax Credit: Corporate Subsidies & Renewable Energy. Heartland Institute. Retrieved April 6, 2023, from

https://www.heartland.org/publications-resources/publications/the-production-tax-credit-corporate-subsidies--renewable-energy.

Essex, C. (2017). Why model? Interface Focus, 7(5), 20170009.

Finkeldey, J. (2018). Challenging the saga of corporate champions. Ephe Journal, 18(4), 817–822.

Fitzgerald, F. (2004). Global warming: a cool view. Ironmaking & Steelmaking, 31, 191 - 198.

Frank, P. (2019). On impact assessment of global mean surface temperature targets at the regional level. Climatic Change, 14, 1–14.

Friedlingstein, P., O'Sullivan, M., Jones, M. W., Andrew, R. M., Hauck, J., Olsen, A., Peters, G. P., Peters, W., Pongratz, J., Sitch, S., Quéré, C. L., Canadell, J. G., Ciais, P., Jackson, R. B., Alin, S., Aragão, L. E. O. C., Arneth, A., Arora, V., Bates, N. R., ... Zaehle, S. Global carbon budget 2020. Earth System Science Data, 12(4), 3269–3340.

https://doi.org/https://doi.org/10.5194/essd-12-3269-2020

Friis-Christensen, E., & Lassen, K. (1994). Solar Activity and Global Temperature. International Astronomical Union Colloquium, 143, 339-347. doi:10.1017/S0252921100024830

Frondel, M., Oertel, K., & Rübbelke, D. (2002). The domino effect in climate change. International Journal of Environment and Pollution, 17, 201.

Funk, C. (2016). The politics of climate. Pew Research Center Science & Society.

https://www.pewresearch.org/science/2016/10/04/the-politics-of-climate/.

Funk, C. (2021, May 26). Key findings: How Americans' attitudes about climate change differ by generation, party and other factors. Pew Research Center.

https://www.pewresearch.org/fact-tank/2021/05/26/key-findings-how-americans-attitudes-about-climate-change-differ-by-generation-party-and-other-factors/

Gale, J.J. (2017). CO 2 Storage in Deep Geological Formations: The Concept.

Gardiner, S. M. (2011). A perfect moral storm: Climate change, intergenerational ethics, and the problem of moral corruption. Environmental Ethics, 33(1), 9-20. doi: 10.5840/enviroethics20113313

Gardiner, S. M. (2011). Ethics and global climate change. Ethics, 114(3), 555-600. doi: 10.1086/660364

Gee, H. (2021, November 30). Humans are doomed to go extinct. Scientific American.

https://www.scientificamerican.com/article/humans-are-doomed-to-go-extinct/

Giddens, A. (2015). The politics of climate change. Policy & Politics, 43(2), 155-162. doi:10.1332/030557315X14290856538163

Goodin, R. E. (2012). Climate change and global public goods: Towards a fairer world order. Oxford University Press.

Green, K. C., & Armstrong, J. S. (2007). On the credibility of climate predictions. Global Environmental Change, 18(3), 397-410.

Green, K.C. and Armstrong, J.S. (2007). Global warming forecasts by scientists versus scientific forecasts. Energy and Environment, 18, 997-1021.

Gronewold, N. (2009, April 2). Who will profit from climate change? Scientific American.

https://www.scientificamerican.com/article/profit-from-climate-change/

Haibach, H., & Schneider, K. (2013). The politics of climate change: Review and future challenges. In O. C. Ruppel, K. B. Gehring, & J. E. Viñuales (Eds.), Climate change: International law and global governance: Volume II: Policy, diplomacy and governance in a changing environment (pp. 357-374). Nomos Verlagsgesellschaft mbH.

https://www.jstor.org/stable/j.ctv941vsk.18

Hamilton, C. (2012). Theories of climate change. Australian Journal of Political Science, 47(4), 721-729. https://doi.org/10.1080/10361146.2 012.732213

Hansen, J., Johnson, D., Lacis, A., Lebedeff, S., Lee, P., Rind, D., & Russell, G. (1981). Climate impact of increasing atmospheric carbon dioxide. Science, 213(4511), 957–966.

https://doi.org/10.1126/science.213.4511.957

Happer, W. (2011). The CO2 theory: A few basic facts. First Things, 218, 35-38. Retrieved from

https://www.firstthings.com/article/2011/02/the-co2-theory-a-few-basic-facts

Harker, D. (2015). Environmental scare: the case of anthropogenic climate change. In Creating Scientific Controversies: Uncertainty and Bias in Science and Society (pp. 175–197). chapter, Cambridge: Cambridge University Press.

Harvey, C. (2016, April 15). Research shows -- yet again -- that there's no scientific debate about climate change. The Washington Post.

https://www.washingtonpost.com/news/energy-environment/wp/2016/04/15/research-shows-yet-again-that-theres-no-scientific-debate-about-climate-change/

Harvey, F., Vaughan, A., & Carrington, D. (2022, November 17). UN chief warns of 'breakdown in trust' with no deal in sight at Cop27. The Guardian.

https://www.theguardian.com/environment/2022/nov/17/un-chief-warns-of-breakdown-in-trust-with-no-deal-in-sight-at-cop27

Heimann. (1993, January 1). The global carbon cycle in the climate system. Springer Berlin Heidelberg. https://link.springer.com/chapter/10.1007/978-3-642-84975-6_9

Held, D., Roger, C., Nag, M., & Shome, P. (2013). The governance of climate change. John Wiley & Sons.

Herman, E. S., & Chomsky, N. (2002). Manufacturing consent: The political economy of the mass media. Pantheon.

Horner, C. C. (2008). Red hot lies: How global warming alarmists use threats, fraud, and deception to keep you misinformed. Simon and Schuster.

Houghton, J. (2009). Global Warming: The Complete Briefing. 4th Edition. Cambridge University Press.

Houghton, J.T., Jenkins, G.J., & Ephraums, J.J. (1990). Climate change: the IPCC scientific assessment.

Hulme, M. (2020). Climates Multiple: Three Baselines, Two Tolerances, One Normal. Academia Letters, Article 102.

https://doi.org/10.20935/AL102.

Humlum, O. (2022). The State of the Climate in 2021. Global Warming Policy Foundation, 51.

Hunt, E. (2021, October 18). The Trick review: How the Climategate scandal rocked the World. New Scientist.

https://www.newscientist.com/article/2294061-the-trick-review-how-the-climategate-scandal-rocked-the-world/

Idso, C.D., Singer, S.F., & Soon, W. (2014). Climate change reconsidered II: Biological impacts. The Heartland Institute.

Isaac, R. J. (2013). Roosters of the Apocalypse—How the junk science of global warming nearly bankrupted the Western world. CreateSpace Independent Publishing Platform.

Jacobson, M. Z. (2009). Short-term effects of controlling fossil-fuel soot, biofuel soot and gases, and methane on climate, Arctic ice, and air pollution health. Journal of Geophysical Research, 114, 1–39.

Jamieson, D. (2014). Reason in a dark time: Why the struggle against climate change failed—and what it means for our future. Oxford University Press.

Janse Van Rensburg, W. (2015). Scepticism about anthropogenic climate disruption: A conceptual exploration [University of Queensland Library]. Retrieved April 6, 2023, from

http://dx.doi.org/10.14264/uql.2015.930

Jensen, N., & Steenhau, I. (2022, August 5). Who, exactly, benefits from renewable energy subsidies? The answer will surprise you. Fast Company.

https://www.fastcompany.com/90776050/who-exactly-benefits-from-renewable-energy-subsidies-the-answer-will-surprise-you

Joffe, J. (2020). The Cult of Climatism: The religion of global warming preaches doom and punishment, even as its own high priests hedge their bets. Meanwhile, its fearful, furious dogmas make a cooperative response to climate change all but impossible. Hoover Digest, (2), 104+. https://link.gale.com/apps/doc/A633468968/AONE?u=anon~64469f97 &sid=bookmark-AONE&xid=e56ba4bf

Johnson, A. (2017). Climate change and global warming - exposed: Hidden evidence, disguised plans. Lulu Press, Inc.

Johnson, C., Affolter, R., Inkenbrandt, S., & Mosher, S. (2019, November 4). 15.5: Anthropogenic Causes of Climate Change. Libretexts. https://geo.libretexts.org/Bookshelves/Geology/Book%3A_An_ Introduction_to_Geology_(Johnson_Affolter_Inkenbrandt_and_ Mosher)/15%3A_Global_Climate_Change/15.05%3A_Anthropogenic_ Causes_of_Climate_Change

Kamarck, E. (2019). The Challenging Politics of Climate Change. Brookings. https://www.brookings.edu/research/the-challenging-politics-of-climate-change/

Kauppinen, J., & Malmi, P. (2019). No experimental evidence for significant anthropogenic climate change. Journal of Geophysical Research, 124(24), 13,602–13,607.

Kearns, L. (2007). Cooking the truth: Faith, science, the market, and global warming. Ecospirit, 97-124. Fordham University Press. http://dx.doi.org/10.5422/fso/9780823227457.003.0006.

Khandekar, M. L., & Pratt, K. F. (2017). Global warming: The citizen's challenge. Science and Public Policy Institute.

Klein, N. (2015). This changes everything: Capitalism vs The climate. Simon and Schuster.

Knight-Jadczyk, L. (2011). Climate change swindlers and the political agenda. Cassiopaea. Retrieved from https://cassiopaea.org/2011/03/21/climate-change-swindlers-and-the-political-agenda/.

Koch, M. (2011). Capitalism and Climate Change: Theoretical Discussion, Historical Development and Policy Responses. Springer.

Kondratenko, T. (2021, June 28). Is global warming merely a natural cycle? Deutsche Welle. https://www.dw.com/en/fact-check-is-global-warming-merely-a-natural-cycle/a-57831350

Kondrat'ev, K. I., Kondratyev, K. Y., Kondrat'ev, K. J., Krapivin, V. F., & Varotsos, C. (2003). Global carbon cycle and climate change. Springer Science & Business Media.

Kramm, G., & Dlugi, R. (2011). Scrutinising the Atmospheric Greenhouse Effect and Its Climatic Impact. Natural Science, 03(12), 971–98.

https://doi.org/10.4236/ns.2011.312124

Langley, P., Munday, M., & Pugh, G. (2017). Exploring the drivers and challenges of demand for the UK Green Deal. Energy Policy, 107, 43-53.

Lear, C.H., Bailey, T.R., Pearson, P.N. et al. (2021). Geological Society of London scientific statement: what the geological record tells us about our present and future climate. Journal of the Geological Society, 178(1), jgs2020-239.

https://doi.org/10.1144/jgs2020-239

Lee, C. C. W., & Lo, Y. T. E. (2015). Effects of aerosol–radiation interaction on precipitation during the East Asian summer monsoon. Atmospheric Chemistry and Physics, 15(2), 771-782.

Legates, D. R., Soon, W., Briggs, W. M., & Monckton, C. (2013). Climate consensus and 'misinformation': A rejoinder to Agnotology, scientific consensus, and the teaching and learning of science. Science & Education, 22(2), 37–71.

Legates, D. R., Soon, W., Briggs, W. M., & Monckton of Brenchley, C. (2013). Climate change: An analysis of IPCC temperature projections. International Journal of Forecasts, 29(3), 669-676. doi: 10.1016/j.ijforecast.2013.01.012

Leveson, I., & Dolan, T. (2016). Green Deal policy in the UK: a critique. Energy Efficiency, 9(2), 329-341.

Lewis, N., & Curry, J. (2018). The impact of recent forcing and ocean heat uptake data on estimates of climate sensitivity. Journal of Climate, 31(15), 6051-6071.

Lewis, N. (2013). An objective Bayesian, improved approach for applying optimal fingerprint techniques to estimate climate sensitivity. Journal of Climate, 26(19), 7414-7429.

Lewis, N. & Crok, M. (2014). Has the climate sensitivity Holy Grail been found? Energy & Environment, 25(6), 1185-1198.

Lewis, N. & Curry, J. (2018). The impact of recent forcing and ocean heat uptake data on estimates of climate sensitivity. Journal of Climate, 31(15), 6051-6071.

Lindzen, R. S., & Choi, Y. S. (2011). On the observational determination of climate sensitivity and its implications. Asia-Pacific Journal of Atmospheric Sciences, 47(4), 377–390.

Lindzen, R. S. (1992). Global warming: The origin and nature of the alleged scientific consensus. Regulation, 15(2), 74-84.

Lindzen, R.S. (1994). CLIMATE DYNAMICS AND GLOBAL CHANGE. Annual Review of Fluid Mechanics, 26, 353-378.

Lindzen, R. S. (2009). How to approach the attribution question? Eos, Transactions American Geophysical Union, 90(13), 111–112.

Lindzen, R. S. (2010). Global warming: A closer look at the numbers. The Wall Street Journal, p. A13. Retrieved from
https://www.wsj.com/articles/SB10001424052748703939404574567423917025400

Lindzen, R. S. (2016). Global Warming and the Irrelevance of Science. In R. Ragaini (Ed.), International Seminars on Nuclear War and Planetary Emergencies 48th Session (pp. 133-143). World Scientific. doi: 10.1142/9789813148994_0012

Lüning, S., Vahrenholt, F., & Crockford, S. J. (2019). Part 1: How reliable are the models? In Climate change: The facts 2019 (pp. 19–35). Stockade Books.

Lüning, S. & Vahrenholt, F. (2019). Evidence for a significant solar effect on climate. Environmental Earth Sciences, 78, 1-18.

MacCracken, M. C. (1983). Climatic effects of atmospheric carbon dioxide. Science, 220(4599), 873-874.
https://doi.org/10.1126/science.220.4599.873

MacFarling Meure, C., Etheridge, D., Trudinger, C., Steele, P., Langenfelds, R., van Ommen, T., Smith, A., & Elkins, J. (2006). Law Dome carbon dioxide, CH4 and N2O ice core records extended to 2000 years BP. Geophysical Research Letters, 33(14), L14810. doi: 10.1029/2006GL026152.

Mann, Geoff. (2018). Climate Leviathan: A Political Theory of Our Planetary Future. Verso Books.

Mann, M. E., Bradley, R. S., & Hughes, M. K. (1999). Northern hemisphere temperatures during the past millennium: Inferences, uncertainties, and limitations. Geophysical Research Letters, 26(6), 759–762.

Manno, J. (2004). Political ideology and conflicting environmental paradigms. Global Environmental Politics, 4(3), 155–159.
https://doi.org/10.1162/1526380041748038

Mark Maslin, The Conversation. (2013, November 14). How climate change and plate tectonics shaped human evolution. Scientific American.
https://www.scientificamerican.com/article/how-climate-change-and-plate-tectonics-shaped-human-evolution/

Marshall, M. (2006, September 4). Timeline: Climate change. New Scientist.
https://www.newscientist.com/article/dn9912-timeline-climate-change/

Maskell, K. (1995). The basic science of anthropogenic climate change. Medicine, Conflict and Survival, 11, 148-167.

Mazza, D., & Canuto, E. (2021b). Evidence of solar 11-year cycle from Sea Surface Temperature (SST). *Academia Letters.*

https://doi.org/10.20935/al3023

Menke, W. (2014). What geology has to say about global warming. Columbia Climate School.

https://news.climate.columbia.edu/2014/07/11/what-geology-has-to-say-about-global-warming/

Metz, B., Davidson, O., Bosch, P., Dave, R., & Meyer, L. (Eds.). (2007). Climate Change 2007: Mitigation of Climate Change. Contribution of Working Group III to the Fourth Assessment Report of the Intergovernmental Panel on Climate Change. Cambridge University Press.

Michaels, P. J., & Balling, R. C. Jr. (2009). Climate of extremes: Global warming science they don't want you to know. Cato Institute.

Michaels, P. J., & Balling Jr, R. C. (2009). Climate of Extremes—Global warming science they don't want you to know. Cato Institute.

Milne, G., & Boardman, B. (2016). The Green Deal and energy company obligation: A critical evaluation. Energy Policy, 95, 270-277.

Monckton, C. (2008). Climate sensitivity reconsidered. Physics and Society, 37(3).

Montford, A. (2015). The Hockey Stick Illusion—Climate and the corruption of science. Stacey International.

Montford, A. (n.d.). Consensus? What consensus? - global warming policy foundation. The Global Warming Policy Foundation. Retrieved November 23, 2022, from

https://www.thegwpf.org/content/uploads/2013/09/Montford-Consensus.pdf

Montford, A. W. (2010). The Hockey Stick Illusion: Climategate and the Corruption of Science. Stacey International Publishers.

Moore, S. (2020, December). The New Geopolitics of Climate Change. The Diplomat.

https://thediplomat.com/2020/12/the-new-geopolitics-of-climate-change/.

Morano, M. (2011, May 31). Special Report: More than 1000 international scientists dissent over man-made global warming claims – challenge UN IPCC & Gore. Climate Depot.

https://www.climatedepot.com/2011/05/31/special-report-more-than-1000-international-scientists-dissent-over-manmade-global-warming-claims-challenge-un-ipcc-gore

Morano, M. (2022). The great reset: Global Elites and the permanent lockdown. Simon and Schuster.

Morrison, I. & Mahowald, N. M. (2019). An observation-based estimate of the strength of rainfall suppression by aerosols in the climate feedback model. Atmospheric Chemistry and Physics, 19(6), 3959-3970.

Mototaka, N. (2018). The fourteen problems of the IPCC models. Natural Science, 10(8), 185–200.

Murray, S. (2021, November 1). How climate change became political. Financial Times.

https://www.ft.com/content/4bac715b-2812-4610-a528-dc8db9ecd635.

Myers, K. F., Maibach, E. W., Roser-Renouf, C., Rosenthal, S. A., Feinberg, G. D., & Leiserowitz, A. A. (2021). Consensus revisited: Quantifying scientific agreement on climate change and climate expertise among Earth scientists 10 years later. Environmental Research Letters, 16(10), 104030. doi: 10.1088/1748-9326/ac2774.

Næss, P. (2012). Climate change and moral responsibility. Journal of Human Rights and the Environment, 3(1), 5-21. doi: 10.4337/jhre.2012.01.02

Narayana, C. N. (2011). Joseph L. Bast and Diane Carol Bast (Eds), Climate Change Reconsidered: The Report of the Nongovernmental International Panel on Climate Change. Chicago, Illinois: The Heart and Institute, 2010, 744 + 12 Pp., Rs 154 (ISBN: 978-1-934791-28-8[PB]). Global Business Review, 12(3), 485-486. doi: 10.1177/097215091101200314.

Nils-Axel Morner, J. (2017). Coastal morphology in relation to sea level changes: A review of concepts and results during the Holocene. Earth-Science Reviews, 176, 53–87.

Nuccitelli, D. (2014, May 28). The Wall Street Journal denies the 97% scientific consensus on human-caused global warming. The Guardian.

https://www.theguardian.com/environment/climate-consensus-97-per-cent/2014/may/28/wall-street-journal-denies-global-warming-consensus.

Nyberg, D., & Wright, C. (2012). Justifying business responses to climate change: Discursive strategies of similarity and difference. Environment and Planning A, 44(8), 1819-1835. doi:10.1068/a45357

Ogawa, S., Kawase, H., Yoshimori, M., Ueda, H., & Oka, A. (2018). Natural variability, radiative forcing and climate response in the recent hiatus reconciled. Scientific Reports, 8(1), 1-8. doi: 10.1038/s41598-018-22458-yL

Oreskes, N. (2004). The Scientific Consensus on Climate Change. Science, 306(5702), 1686-1686. doi: 10.1126/science.1103618.

Oreskes, N. (2018). The scientific consensus on climate change: How do we know we're not wrong? In Climate Modelling (pp. 31-64). Springer International Publishing. doi: 10.1007/978-3-319-65058-6_2.

Oschlies, A., Rehder, G., Kopf, Achim, Riebesell, Ulf, Wallmann, K., & Zimmer, M. (2023). The Earth's natural carbon cycle – Carbon reservoir ocean: How the sea absorbs carbon dioxide. CDRmare.

http://dx.doi.org/10.3289/cdrmare.24

Overland, I., & Sovacool, B. K. (2020). The Misallocation of Climate Research Funding. Energy Research & Social Science, 62, 101349.

https://doi.org/10.1016/j.erss.2019.101349

Overland, I. (2019). The geopolitics of renewable energy: Debunking four emerging myths. Energy Research & Social Science, 49, 36-40. doi: 10.1016/j.erss.2018.10.018.

O'Neill, S., & Nicholson-Cole, S. (2009). 'Fear Won't Do It.' Science Communication, 30(3), 355–379.

https://doi.org/10.1177/1075547008329201

Page, M. L., & Brahic, C. (2007, May 16). Climate myths: Ice cores show CO2 increases lag behind temperature rises, disproving the link to global warming. New Scientist.

https://www.newscientist.com/article/dn11659-climate-myths-ice-cores-show-co2-increases-lag-behind-temperature-rises-disproving-the-link-to-global-warming/

Palmer, J., & Cooper, I. (2017). The UK's Green Deal: a case study in policy failure. Environmental Politics, 26(5), 814-834.

Panofsky, H. A. (1956). Theories of climate change. Weatherwise, 9, 183-204.

Patterson, A. (2022). Climate Catastrophe: The Role of Fear Appeals in Climate Change Communication. Victoria University of Wellington Library.

http://dx.doi.org/10.26686/wgtn.19446845.

Petersen, A.C. (2000). Philosophy of climate science. Bulletin of the American Meteorological Society, 81, 265-271.

Pielke, R. A. (2013). The scientific basis for human-induced climate change. Environmental Science & Policy, 26, 1–15.

Pielke Sr, R. A. (2008). A broader view of the role of humans in the climate system. Physics Today, 61(9), p. 54–55.

Pilkey, O. H., Pilkey-Jarvis, L., & Young, R. S. (2011). Global Climate Change. Duke University Press.

Plimer, I. (2007). Climate change delusion and the great electricity ripoff.

Plimer, I. (2009). Heaven and Earth: Global Warming, the Missing Science. Connor Court.

Plimer, I. (2017). Climate Change Delusion and the Great Electricity Rip-Off. Connor Court.

Plimer, I. R. (2009). Heaven and Earth: Global Warming, the Missing Science. Connor Court.

Plimer, I.R. (2009). Heaven and Earth: Global Warming, the Missing Science. Taylor Trade Publishing.

Popescu, G. N., Popescu, V. A., & Popescu, C. R. (2014b). Corporate governance in Romania: Theories and practices. In Corporate Governance and Corporate Social Responsibility (pp. 375–401). WORLD SCIENTIFIC.
http://dx.doi.org/10.1142/9789814520386_0014

Powell, J. L. (2016). The Consensus on Anthropogenic Global Warming Matters. Bulletin of Science, Technology & Society, 36(3), 157-163.
https://doi.org/10.1177/0270467617707079.

Pralle, S. B. (2009). Agenda-Setting and Climate Change. Environmental Politics, 18(5), 781-799. doi:10.1080/09644010903157115.

Priem, H.N. (1997). Co2 and Climate: a Geologist's View. Space Science Reviews, 81, 173-198.

Rajan, S. (2017). Abrupt Climate Shifts Over the Past 10,000+ Years: An Arctic-Antarctic-Asian Imbroglio? In Science and Geopolitics of the White World (pp. 83-91). Springer International Publishing.
http://dx.doi.org/10.1007/978-3-319-57765-4_7.

Reichle, D. E. (2023). The global carbon cycle and climate change: Scaling ecological energetics from organism to the biosphere. Elsevier.

Reinhart, A., et al. (2018). Climate engineering reconsidered. Nature Climate Change, 8, 209–217.

Reusswig, F. (2013). History and Future of the Scientific Consensus on Anthropogenic Global Warming. Environmental Research Letters, 8(3), 031003. doi:10.1088/1748-9326/8/3/031003.

Roskam, J., & Moran, A. J. (2010). Climate Change: The Facts. Institute of Public Affairs.

Rotty, R. M. (1979). Uncertainties associated with global effects of atmospheric carbon dioxide. Office of Scientific and Technical Information (OSTI). doi: 10.2172/6291158

Ruedy, R., & Hansen, J. E. (2019). Confronting atmospheric greenhouse gases: Three decades of global climate change. Atmosphere, 10(5), 1–19.

Schiermeier, Q. The real holes in climate science. Nature 463, 284–287 (2010).
https://doi.org/10.1038/463284aOreskes,

Schlosberg, D. (2013). Climate justice and capabilities: A framework for adaptation policy. Ethics & International Affairs, 27(4), 445-461. doi: 10.1017/S0892679413000262

Schulte, K. (2008). Scientific consensus on climate change? Energy & Environment, 19(2), 281-286. doi: 10.1260/095830508783900744.

Scruton, R. (2010). Green philosophy. Atlantic Monthly Press.

Sharon, O. (2021). State extinction through climate change. In Debating Climate Law (pp. 349-364). Cambridge University Press. doi: 10.1017/9781108879064.026.

Shaviv, N. J. (2005). On climate response to changes in the cosmic ray flux and radiative budget. Journal of Geophysical Research, 110, D10108.

Singer, S. F. (2007, May 22). The Great Global Warming Swindle. The Independent Institute.

https://www.independent.org/news/article.asp?id=1945.

Singer, S. F. (2008). Global warming and climate change: Science and politics. Civil Engineering, 78(2), 38-47. Retrieved from

https://www.researchgate.net/publication/23436087_Global_warming_and_climate_change_Science_and_politics

Singh, S., Kulshrestha, U., Aggarwal, S., & Singh, S. (2021). Global Climate Change. Elsevier.

Smith, I.M. (1993). CO2 and climatic change: An overview of the science. Energy Conversion and Management, 34, 729-735.

Smith, T. M. (1998). The myth of green marketing: Tending our goats at the edge of apocalypse. University of Toronto Press.

Solomon, L. (2010). The Deniers: The World-Renowned Scientists Who Stood Up Against Global Warming Hysteria, Political Persecution, and Fraud. Richard Vigilante Books.

Solomon, S., Qin, D., Manning, M., Chen, Z., Marquis, M., Averyt, K. B., ... & Tignor, M. (2007). Climate change 2007: The physical science basis. Cambridge University Press.

Soon, W., & Baliunas, S. L. (2003). Global warming. Progress in Physical Geography, 27, 448-455.

Soon, W., Baliunas, S., Idso, S. B., Kondratyev, K. Ya., Posmentier, E. S., & Sallie, L. W. (1999). Environmental effects of increased atmospheric carbon dioxide. Climate Research, 13(2), 149-164.

https://doi.org/10.3354/cr013149

Soon, W., Legates, D. R., Briggs, W. M., & Monckton, C. (2015). Climate modelling: inaccurate or dishonest? Journal of Scientific Exploration, 29(3), 441–476.

Spencer, R. W., & Braswell, W. D. (2014). The role of ENSO in global climate change in the context of the current warming hiatus. Climate Dynamics, 43(1), 1–15.

Spencer, R. W. (2017). Temperature trends in the lower atmosphere: Steps for understanding and reconciling differences. Current Climate Change Reports, 3(4), 242–254.

Spencer, R.W. (2017). Tropospheric temperature trends: history of ongoing controversy. Wiley Interdisciplinary Reviews: Climate Change, 8(4). doi: 10.1002/wcc.462

Spencer, R. W. (2019). The misdiagnosis of surface temperature feedbacks from variations in Earth's radiant energy balance. Remote Sensing, 11(8), 1–26.

Sperling, D., & Cannon, J. S. (2010). Driving Climate Change: Cutting Carbon from Transportation. Elsevier.

Stacy, T., & Taylor, G. (n.d.). Publications - The Levelized Cost of Electricity from Existing Generation Resources. Heartland Institute. Retrieved from

https://www.heartland.org/publications-resources/publications/the-levelized-cost-of-electricity-from-existing-generation-resources.

Stecuła, D. (2017, August 16). An Inconvenient Truth. The Conversation.

https://theconversation.com/an-inconvenient-truth-about-an-inconvenieSummerhayes, C. P., Maslin, M. A., & Day, S. J. (2006). The geological perspective of global warming. Global Warming Policy Foundation.

https://www.thegwpf.org/content/uploads/2013/12/Geological-Perspective.pdf.

Steffen, W., Broadgate, W., Deutsch, L., Gaffney

Sun, Y., & Shi, B. (2022). Impact of greenwashing perception on consumers' green purchasing intentions: A moderated mediation model. Sustainability, 14(19), 12119.

https://doi.org/10.3390/su141912119

Sussman, B. (2010). Climategate: A veteran meteorologist exposes the global warming scam. WND Books.

Svensmark, H. (2014). Evidence for global warming due to increasing solar radiation. Advances in Space Research, 53(12), 1452–1459.

Svensmark, H. & Calder, N. (2006). The chilling stars: A new theory of climate change. Icon Books.

Syvitski, J., Waters, C., Day, J., Milliman, J. D., Summerhayes, C., Steffen, W., Zalasiewicz, J., Cearreta, A., Galuszka, A., Hajdas, I., Head, M. J., Leinfelder, R., McNeill, J. R., Poirier, C., Rose, N. L., Shotyk, W., Wagreich, M., & Williams, M. (2020). Extraordinary human energy consumption and resultant geological impacts beginning around 1950

CE initiated the proposed Anthropocene Epoch. Communications Earth & Environment, 1, 32.

https://doi. org/10.1038/s43247-020-00029-y

Tait, N., & Waddams, C. (2015). The UK Green Deal: A model for financing energy efficiency? Energy Policy, 84, 220-232.

Taylor, J. (2013, May 30). Global Warming Alarmists Caught Doctoring '97-Percent Consensus' Claims. Forbes.

https://www.forbes.com/sites/jamestaylor/2013/05/30/global-warming-alarmists-caught-doctoring-97-percent-consensus-claims/.

The Editors of Encyclopaedia Britannica, Patrick Riley. (2021). Timeline of climate change. Encyclopedia Britannica. Retrieved February 26, 2023, from

https://www.britannica.com/story/timeline-of-climate-change

The Royal Society and National Academy of Sciences. (2014). Climate Change: Evidence and Causes. National Academies Press.

Todaro, N., & Li, H. (2016). The UK Green Deal: a new paradigm for energy efficiency in the residential sector?. Energy Efficiency, 9(2), 343-356.

Tol, R. (2014, June 6). The claim of a 97% consensus on global warming does not stand up. The Guardian.

https://www.theguardian.com/environment/blog/2014/jun/06/97-consensus-global-warming

Tsonis, A. A., Swanson, K. L., & Kravtsov, S. (2007). A new dynamical mechanism for major climate shifts. Geophysical Research Letters, 34(13), L13705.

Uscinski, J. E., & Olivella, S. (2017). The Conditional Effect of Conspiracy Thinking on Attitudes toward Climate Change. Research & Politics, 4(4), 205316801774310.

https://doi.org/10.1177/2053168017743105.

Uscinski, J.E., Klofstad, C.A., Atkinson, M.D., & McDermott, R. (2017). Climate Change Conspiracy Theories. Oxford Research Encyclopedia of Climate Science. Oxford University Press.

http://dx.doi.org/10.1093/acrefore/9780190228620.013.328.

Van de Graaf, T., Sovacool, B.K., Andrews, N., and van der Laan, G. (2020). The new oil? The geopolitics and international governance of hydrogen. Energy Research & Social Science, 70, 101667. doi: 10.1016/j.erss.2020.101667.

Vinod, D. (2020, August 12). Climate change through the Geological Era. Deepchand Vinod - Academia.Edu.

https://www.academia.edu/43841005/Climate_change_through_Geological_Era?email_work_card=title

Vinod, D. (2020, August 12). Climate change through the Geological Era. Deepchand Vinod - Academia.Edu.

https://www.academia.edu/43841005/Climate_change_through_Geological_Era?email_work_card=title

Von Storch, H. (2012). Anthropogenic Climate Change. Springer.

Voosen, P. (2019a, October 30). Science. AAAS.

https://www.science.org/content/article/world-s-oldest-ice-core-could-solve-mystery-flipped-ice-age-cycles

Watson, J., & Gross, R. (2014). The UK's Green Deal: A tentative step towards a more sustainable energy policy?. Energy Policy, 66, 419-426.

Weber, E. U. (2010). What Shapes Perceptions of Climate Change?. Wiley Interdisciplinary Reviews: Climate Change, 1(3), 332–342.

https://doi.org/10.1002/wcc.41.

Will. (2019c, February 23). Climate change and the ten warning signs for cults - Will. Medium.

https://medium.com/@hwater84/climate-change-and-the-ten-warning-signs-for-cults-56c181db82c1

World Resources Institute. (2010). Climate Analysis Indicators Tool (CAIT). Retrieved from http://cait.wri.org/

Wright, C., & Nyberg, D. (2015). Climate Change, Capitalism, and Corporations: Processes of Creative Self-Destruction. Cambridge University Press.

Young, L.D., & Fitz, E.B. (2021). Who are the 3 per cent? The connections among climate change contrarians. British Journal of Political Science, 52(4), 1503-1522. doi:10.1017/s0007123421000442

Zalasiewicz, J., Waters, C., Williams, M., & Summerhayes, C. (Eds.). (2019). The Anthropocene as a Geological Time Unit: A Guide to the Scientific Evidence and Current Debate. Cambridge: Cambridge University Press. doi:10.1017/9781108621359

Zalasiewicz, J., Williams, M., Haywood, A., & Ellis, M. (2011). The Anthropocene: a new epoch of geological time? Philosophical Transactions, 369, 835–841.

https://doi.org/10.1098/rsta.2010.0339

Zhu, Z., Piao, S., Myneni, R.B., et al. (2016). "Greening of the Earth and its drivers." Nature Climate Change, 6, 791-795. DOI: 10.1038/nclimate3004. Available at:

https://www.nature.com/articles/nclimate3004

About the Author

Balvinder Ruby, the author of "Climate Conundrum - The Agendas and the Forces at Play", is an expert in the field of earth science who has worked as a geoscientist for over two decades. With two published books under his belt, "The Provocateur - Spilling Beans by All Means" and "New World Order - The Rise of Transnational Corporate Republic", the author is a seasoned writer.

Apart from being an accomplished writer, the author is also a brand ambassador and contributor to The Times of India newspaper from Sydney. Through his writings, the author aims to create awareness and educate people about the complexities of climate change and the agendas and forces at play behind it.

With his rich experience and knowledge, the author provides a unique perspective on the subject matter of the book. His insights will undoubtedly enlighten and inspire readers to take a more active role in addressing the critical issue of climate change.

You can connect with him at
https://balvinderruby.com.au

Books by the same Author

The Provocateur: Spilling beans by all means

It is a collection of 38 poems that aim to motivate, inspire, and uplift readers. Each poem is centred around a specific theme and is accompanied by black-and-white pencil sketches.

The book's goal is to guide readers through their own thoughts and emotions, encouraging them to explore their potential and approach daily life with harmony.

By challenging readers to step out of their comfort zones, confront their obstacles, and take charge of their lives, the book urges them to embrace their true selves, stand up for their rights, and make their own decisions about the future.

What others say

"an extraordinary piece of writing, intelligent and very reflective"

– Adjunct Professor Dr Jim Taggart, OAM,
Deputy Chairman Riverside Theatres

New World Order: The Rise of the Transnational Corporate Republic

This book analyses the factors that led to the Cold War between the liberal, democratic capitalist bloc led by the United States and the communist, socialist bloc led by the former Union of Soviet Socialist Republics (USSR) after the two world wars. NATO and the Warsaw Pact were military and ideological competitors that faced off against one another to keep a check on each other's strengths. With the Soviet Union's demise, the United States became the world's lone arbiter in terms of geopolitical and economic stability.

China's rise to prominence as the world's industrial hub may be traced back to the United States and other Western countries' conception and promotion of globalisation to access emerging markets. During the COVID-19 epidemic, when the global supply chain from China broke down, it became painfully clear that almost all manufactured goods around the world came from China.

Because of outsourcing, which was made possible by developments in IT and telecom, the services sector has moved to developing countries like India. As a result of China's economic rise, the West and the United States have seen their dominance erode due to a loss of industrial and service sector jobs.

Additionally, the idea of nation-state sovereignty and democratic rule became watered down due to globalisation. The result was the emergence of international companies and the corporatisation of democracies.

Understanding how the dynamics of the fall of the democratic model of governments, corporatisation, and the rise of China will play out in defining the developing new international order is the goal of this book.

www.ingramcontent.com/pod-product-compliance
Lightning Source LLC
Chambersburg PA
CBHW031124180526
45160CB00006B/69/J